BIG IDEAS MATH®
Algebra 1
A Common Core Curriculum

Assessment Book

- Pre-Course Test with Item Analysis
- Quizzes
- Chapter Tests
- Standards Assessment with Item Analysis
- Alternative Assessments
- End-of-Course Tests

Erie, Pennsylvania

Copyright © by Big Ideas Learning, LLC. All rights reserved.

Permission is hereby granted to teachers to reprint or photocopy in classroom quantities only the pages or sheets in this work that carry a Big Ideas Learning copyright notice, provided each copy made shows the copyright notice. These pages are designed to be reproduced by teachers for use in their classes with accompanying Big Ideas Learning material, provided each copy made shows the copyright notice. Such copies may not be sold and further distribution is expressly prohibited. Except as authorized above, prior written permission must be obtained from Big Ideas Learning, LLC to reproduce or transmit this work or portions thereof in any other form or by any other electronic or mechanical means, including but not limited to photocopying and recording, or by any information storage or retrieval system, unless expressly permitted by copyright law. Address inquiries to Permissions, Big Ideas Learning, LLC, 1762 Norcross Road, Erie, PA 16510.

Big Ideas Learning and *Big Ideas Math* are registered trademarks of Larson Texts, Inc.

ISBN 13: 978-1-60840-473-5
ISBN 10: 1-60840-473-0

123456789-VLP-17 16 15 14 13

Contents

About the Assessment Book iv

Pre-Course Test with Item Analysis 1

Chapter 1 Solving Linear Equations 5

Chapter 2 Graphing and Writing Linear Equations 17

Chapter 3 Solving Linear Inequalities 29

Chapter 4 Solving Systems of Linear Equations 41

Chapter 5 Linear Functions 53

Chapter 6 Exponential Equations and Functions 65

Chapter 7 Polynomial Equations and Factoring 77

Chapter 8 Graphing Quadratic Functions 89

Chapter 9 Solving Quadratic Equations 101

Chapter 10 Square Root Functions and Geometry 113

Chapter 11 Rational Equations and Functions 125

Chapter 12 Data Analysis and Displays 137

End-of-Course Tests 149

Gridded Response Answer Sheet 157

Answers A1

About the Assessment Book

Pre-Course Test with Item Analysis

The Pre-Course Test covers material that students should be familiar with from earlier courses. The Item Analysis can be used to determine topics that need to be reviewed.

Quizzes

The Quizzes provide ongoing assessment of student understanding. There are two quizzes for each chapter.

Chapter Tests

The Chapter Tests provide assessment of student understanding of key concepts taught in the chapter. There are two tests for each chapter.

Standards Assessment with Item Analysis

The Standards Assessment provides students practice answering questions in state assessment format. The assessments are cumulative and cover material from the current chapter as well as earlier chapters of the textbook. Questions are presented in multiple choice, gridded response, short response, and extended response format. The Item Analysis can be used to identify common errors and assess student understanding.

Alternative Assessment with Scoring Rubric

Each Alternative Assessment includes at least one multi-step problem that combines a variety of concepts from the chapter. Students are asked to explain their solutions, write about the mathematics, or compare and analyze different situations.

End-of-Course Tests

The End-of-Course Tests cover the key concepts taught throughout the course and can be used as a year-end exam or as a practice test to help students prepare for state assessments.

Gridded Response Answer Sheet

The Gridded Response Answer Sheets can be used to help students practice completing gridded response questions.

Pre-Course Test

Tell whether the two fractions form a proportion.

1. $\dfrac{3}{4}, \dfrac{16}{20}$
2. $\dfrac{5}{7}, \dfrac{30}{42}$
3. $\dfrac{4}{18}, \dfrac{6}{27}$

4. You buy a sweater that is discounted 25%. The original price of the sweater is $40. What is the price of the sweater after the discount?

5. Find the actual distance between Lisbon and Barcelona.

Find the coordinates of the point.

6. A
7. B

Plot the ordered pair.

8. $(1, 4)$
9. $(3, -2)$

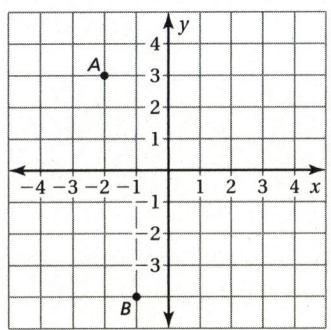

Simplify the expression.

10. $-4 + 11$
11. $-7(-8)$
12. $60 \div (-4)$
13. $|-34|$
14. $|-(-41)|$
15. $12 - (-19)$
16. $\dfrac{4}{15} + \dfrac{5}{9}$
17. $-\dfrac{7}{8} \div \dfrac{3}{4}$
18. $\dfrac{13}{18} \cdot \dfrac{9}{25}$
19. $-\dfrac{7}{12} - \dfrac{1}{8}$
20. $8.37(-5.3)$
21. $0.95 - 3.49$

Answers

1. _____
2. _____
3. _____
4. _____
5. _____
6. _____
7. _____
8. __See left.__
9. __See left.__
10. _____
11. _____
12. _____
13. _____
14. _____
15. _____
16. _____
17. _____
18. _____
19. _____
20. _____
21. _____

Pre-Course Test (continued)

Solve the equation, if possible.

22. $x - 9 = -2$

23. $-4x = 32$

24. $9 - 2x = 23$

25. $x - 7 = x + 6$

26. $4x - 2 = x - 5$

27. $4x + 12 = 4(3 + x)$

28. Use the properties of equality to show that the equation $6x + 3 = 27$ is equivalent to the equation $2x = 8$.

Find the side length of the square.

29.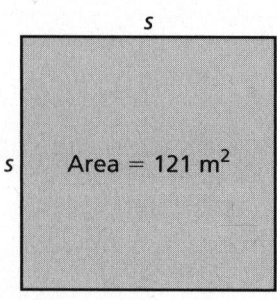
Area = 121 m²

30.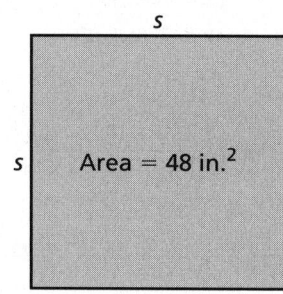
Area = 48 in.²

Write the fraction as a decimal.

31. $\dfrac{3}{4}$

32. $\dfrac{5}{16}$

33. $\dfrac{21}{4}$

34. In a class, the teacher asks each person wearing red to name his or her favorite color. Is this sample representative of the entire class? Explain.

35. The data below are the test scores of the students in a math class.

    97, 76, 84, 82, 90, 95, 77, 79, 80, 82, 84, 77, 100, 78, 87

    Create a stem-and-leaf plot to represent the data.

36. Find the slope and y-intercept of the graph of $y = 3x - 8$.

Simplify the expression.

37. $\sqrt{25}$

38. $\sqrt[3]{-8}$

39. $\sqrt{54}$

40. $(-6)^2$

41. $(3d)^4$

42. $\dfrac{2^3}{2^5}$

Answers

22. _____
23. _____
24. _____
25. _____
26. _____
27. _____
28. _____
_____
_____
29. _____
30. _____
31. _____
32. _____
33. _____
34. _____
_____
_____
35. ___See left.___
36. _____
_____
_____
37. _____
38. _____
39. _____
40. _____
41. _____
42. _____

Pre-Course Test Item Analysis

Item Number	Skills
1	simplifying fractions, understanding proportion
2	simplifying fractions, understanding proportion
3	simplifying fractions, understanding proportion
4	solving a problem involving a discount
5	understanding scale
6	identifying the coordinates of a point
7	identifying the coordinates of a point
8	plotting a point in a coordinate plane
9	plotting a point in a coordinate plane
10	adding integers
11	multiplying integers
12	dividing integers
13	finding absolute value of integers
14	finding absolute value of integers
15	subtracting integers
16	adding fractions
17	dividing fractions
18	multiplying fractions
19	subtracting fractions
20	multiplying decimals
21	subtracting decimals

Item Number	Skills
22	solving one-step equations
23	solving one-step equations
24	solving two-step equations
25	recognizing equations with no solution
26	solving equations with variables on both sides
27	solving equations with infinitely many solutions
28	using properties of equality
29	finding a square root
30	simplifying a square root
31	writing a fraction as a decimal
32	writing a fraction as a decimal
33	writing a fraction as a decimal
34	evaluating a sample
35	making a stem-and-leaf plot
36	finding the slope and y-intercept of a linear function
37	finding a square root
38	finding a cube root
39	simplifying a square root
40	raising a number to an exponent
41	using properties of exponents
42	using properties of exponents

Name_____ Date_____

Chapter 1 Quiz
For use after Section 1.2

Solve the equation. Check your solution.

1. $4 - c = -\dfrac{1}{3}$

2. $-14 = x - 12$

3. $\dfrac{s}{1.5} = 0.8$

4. $0.4r = 1.6$

Solve the equation. Check your solution.

5. $2d - 15 = 3$

6. $4 = \dfrac{3}{4}m - 6$

7. $-3(n + 6) + 10 = -8$

8. $1.5(q - 4) - 2 = 4$

Find the value of x.

9.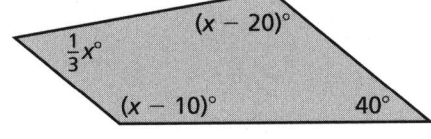

Sum of angle measures: 360°

10.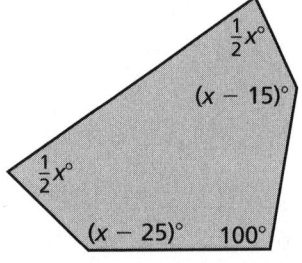

Sum of angle measures: 540°

11. The equation $R = 10A - 20$ represents the revenue R (in dollars) you make by spending A dollars on advertising. Your revenue totaled $110. How much did you spend on advertising?

12. A 150-foot fence encloses a garden. What is the length of each side of the garden?

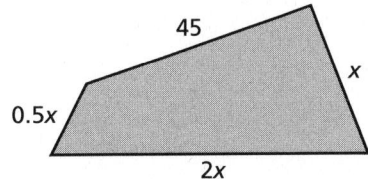

13. A car drives 60 miles per hour. Write and solve an equation to find the number of hours it takes the car to travel 360 miles.

14. Use the table to write and solve an equation to find the number of cars c that a salesperson needs to sell in the fourth month so that the salesperson's mean number of cars sold per month is 10.

Month	1	2	3	4
Cars Sold	8	9	12	c

Answers

1. _____
2. _____
3. _____
4. _____
5. _____
6. _____
7. _____
8. _____
9. _____
10. _____
11. _____
12. _____
    _____
13. _____
    _____
14. _____
    _____

Chapter 1 Quiz

For use after Section 1.4

Solve the equation. Check your solution.

1. $-4p = 3p + 28$
2. $-2y - 4 = 4(y - 1)$

Solve the equation. Graph the solutions, if possible.

3. $|k + 5| = 7$
4. $-2|z + 8| = -6$

Solve the equation for y.

5. $5x - 4y = 10$
6. $7 = -y + 3x$

7. The formula for the volume V of a cone is $V = \frac{1}{3}\pi r^2 h$. Solve the formula for the height h.

8. The formula for the area A of a triangle is $A = \frac{1}{2}bh$. Solve the formula for the base length b.

9. It is 35°C at your school and 90°F at home. Where is the temperature higher?

10. The area of a trapezoid is $A = \frac{1}{2}h(b + B)$. Solve the formula for the height h. What is the height if the area is 200 square feet, the length of the smaller base b is 10 feet, and the length of the larger base B is 15 feet?

11. From your home, the route to the school that passes the mall is 2 miles shorter than the route to the school that passes the theater. What is the length of each route?

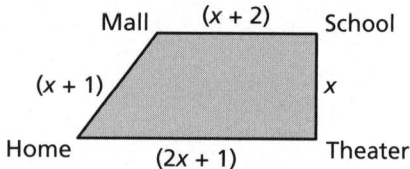

12. The formula for the surface area S of a cylinder is $S = 2\pi r^2 + 2\pi rh$. Solve the formula for the height h.

Answers

1. _____
2. _____
3. _____
   See left.
4. _____
   See left.
5. _____
6. _____
7. _____
8. _____
9. _____
10. _____
    _____
11. _____
    _____
12. _____

Name _____ Date _____

Chapter 1 Test A

Solve the equation. Check your solution.

1. $y - 12 = 9$
2. $0.6 = r + 4.2$
3. $42 = 7x$
4. $\dfrac{f}{5} = -4$
5. $5p - 7 = 28$
6. $22 - 6g = 18$
7. $1.5x + 1.3x = -8.4$
8. $5r + 8 = 2r$
9. $\dfrac{4}{3}w - 12 = \dfrac{2}{3}w$
10. $4(3q - 2) = 16q$

Solve the equation. Graph the solutions, if possible.

11. $|2d - 5| = 3$

12. $|3t + 9| = 6$

Solve the equation for y.

13. $\dfrac{2}{5}x + y = 3$
14. $8 = 3x + 6y$
15. $1.5x - 3y = 6$
16. $\dfrac{1}{4}y - 2x = 5$

17. The formula for profit is $P = R - C$.

    a. Solve the formula for R.

    b. Use the new formula to find the value of R given that $P = \$350$ and $C = \$520$.

Solve the equation for the bold variable.

18. $V = \ell w \mathbf{h}$
19. $s = \mathbf{p} - 0.2t$
20. $Z = s\mathbf{L}$
21. $\mathbf{P}V = nRT$

Answers

1. _____
2. _____
3. _____
4. _____
5. _____
6. _____
7. _____
8. _____
9. _____
10. _____
11. _____
   See left.
12. _____
   See left.
13. _____
14. _____
15. _____
16. _____
17. a. _____
    b. _____
18. _____
19. _____
20. _____
21. _____

Chapter 1 Test A (continued)

22. There are 24 more students in the seventh grade class than the number g in the eighth grade class. The seventh grade class has 160 students. Write and solve an equation to find the number of students in the eighth grade class.

23. You rent a canoe for $5 per hour. Your cost before the tax is added is $12.50. Write and solve an equation to find the number of hours that you rented the canoe.

24. The cost (in dollars) of making n birthday cakes is represented by $C = 24n + 35$. How many birthday cakes are made when the cost is $395? Explain your reasoning.

Answers

22. _____

_____

23. _____

_____

24. ___See left.___

25. _____

25. The sum of the interior angle measures is 180°. Find the value of x.

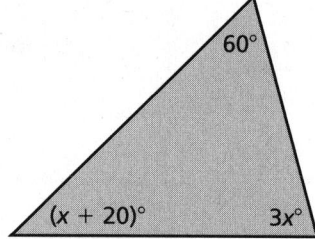

Name_____ Date_____

Chapter 1 Test B

Solve the equation. Check your solution.

1. $-26d = -364$

2. $x - \dfrac{1}{3} = \dfrac{2}{5}$

3. $\dfrac{5}{6} = -\dfrac{2}{3}n$

4. $4\pi = s + 5\pi$

5. $1.4x + 1.2x - 3.8 = -9$

6. $\dfrac{3}{5}w - \dfrac{1}{5}w + 10 = 4$

7. $4(3 - 6a) = 36$

8. $4(2g - 3) = 5(g - 2)$

9. $25(1 - x) = -8x - 9$

10. $0.6(j - 2) = 0.3j$

Solve the equation. Graph the solutions, if possible.

11. $-3|k - 5| = -12$

12. $2|c + 2| = -6$

Solve the equation for y.

13. $2\pi = 5x - 3y$

14. $2.4x - 1.5y = 3$

15. $2.7 = 5.4y - 8.1x$

16. $\dfrac{1}{3}x + \dfrac{2}{3}y = 1$

17. The formula for simple interest is $I = Prt$.

   a. Solve the formula for P.

   b. Use the new formula to find the value of P in the table.

I	$40
P	?
r	4%
t	2

Solve the equation for the bold variable.

18. $i = \dfrac{e\mathbf{c}}{3}$

19. $P = 2\ell + 2\mathbf{w}$

20. $I = \dfrac{V}{\mathbf{R}}$

21. $S = 3\pi r^2 + 2\pi r\mathbf{h}$

Answers

1. _____
2. _____
3. _____
4. _____
5. _____
6. _____
7. _____
8. _____
9. _____
10. _____
11. _____
   __See left.__
12. _____
   __See left.__
13. _____
14. _____
15. _____
16. _____
17. a._____
    b._____
18. _____
19. _____
20. _____
21. _____

Name _____ Date _____

Chapter 1 Test B (continued)

22. The cost of your new book bag is $11.50 more than the cost c of your old book bag. You pay $47 for your new book bag. Write and solve an equation to find the cost of your old book bag.

23. You can rent a video game for $3.50. Your total cost of rentals for the month was $31.50. Write and solve an equation to find the number of video game rentals for the month.

24. You purchase 5 movies and a CD. The cost of the CD is $8.50. Your total bill before tax is $38.45. Write and solve an equation to find the cost of a movie.

25. The formula $C = p + 0.06p$ represents the after-tax cost C for an item with a purchase price p.

  a. Solve for p.

  b. Find the purchase price if the after-tax cost is $62.54.

26. The sum of the interior angles of the quadrilateral is 360°. Find the value of x.

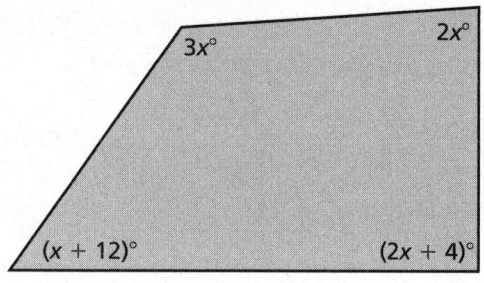

Answers

22. _____

_____

23. _____

_____

24. _____

_____

25. a. _____

b. _____

26. _____

Name_____ Date_____

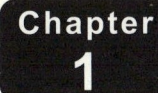 **Standards Assessment**

1. **GRIDDED RESPONSE** The square shown below has a perimeter of $6x - 20$ units. What is the value of x?

   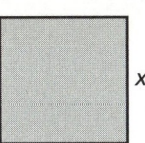

2. The formula $PV = 300R$ is used in chemistry. How can this formula be solved for P?

   F. Divide both sides of the formula by P.

   G. Divide both sides of the formula by V.

   H. Subtract P from both sides of the formula.

   I. Subtract V from both sides of the formula.

3. Emma received a $40 gift card that could be used to download television programs. After she downloaded 5 programs, she had $30 remaining on her gift card. If each program costs the same amount to download, what is the cost of one download?

   A. $2 **C.** $8

   B. $6 **D.** $10

4. The steps Christa took to solve the equation are shown below. What should Christa change in order to correctly solve the equation?

   $$|x - 4| + 3 = 8$$
   $$x - 4 + 3 = 8 \quad \text{or} \quad x - 4 + 3 = -8$$
   $$x - 1 = 8 \quad\quad\quad\quad x - 1 = -8$$
   $$x = 9 \quad \text{or} \quad\quad\quad x = -7$$

   A. Nothing; she solved the equation correctly.

   B. Remove the absolute value from the original equation and solve the resulting linear equation.

   C. Subtract 3 from both sides before writing two separate linear equations.

   D. Subtract 3 from both sides and then add 4 to both sides before writing two separate linear equations.

Chapter 1 Standards Assessment (continued)

5. The formula for the area of a trapezoid is $A = \dfrac{h}{2}(a + b)$. Solve the equation for h.

   F. $h = 2A(a + b)$ **H.** $h = A - 2a - 2b$

   G. $h = \dfrac{2A}{a + b}$ **I.** $h = \dfrac{A}{2(a + b)}$

6. What value of x makes the equation below true?

   $$5x + 9 = x + 20$$

   A. 7.25   **C.** 2.75

   B. 5.8   **D.** 2.2

7. The drawing shows equal weights on two sides of a balance scale.

   The can of coffee weighs 20 ounces. Each book weighs 36 ounces. What is the weight, in ounces, of one carton of salt?

   F. 8   **H.** 46

   G. 26   **I.** 52

8. **EXTENDED RESPONSE** Cody is hiring a lawyer to help him collect a debt.

   Part A Lawyer A charges an hourly rate. She estimates the case will take 20 hours and cost $4900. Write and solve an equation to determine her hourly rate x.

   Part B Lawyer B charges an hourly rate and an up-front fee of $650. He estimates the case will take 22 hours and cost $5490. Write and solve an equation to determine his hourly rate y.

   Part C Use the hourly rates calculated in parts A and B to write and solve an equation to determine the hours t of work the case would have to take for Cody to get the same price from each lawyer.

Chapter 1 Standards Assessment Item Analysis

1. Correct answer: 10

   Common error: The student confuses or does not know the meaning of perimeter, and sets x equal to $6x - 20$, yielding the answer $x = 4$.

2. **A.** The student misunderstands what it means to solve for P.
   B. Correct answer
   C. The student misunderstands what it means to solve for P and thinks that PV can be separated using subtraction.
   D. The student thinks that PV can be separated using subtraction instead of division.

3. **F.** Correct answer
   G. The student divides $30 by 5.
   H. The student divides $40 by 5.
   I. The student subtracts the two dollar amounts given in the problem, $40 − $30.

4. **A.** The student doesn't solve for the absolute value part before writing two related linear equations.
   B. The student disregards the meaning of absolute value.
   C. Correct answer
   D. The student misunderstands the process of solving an absolute value equation by solving two related linear equations.

5. **F.** The student multiplies the left side of the equation by $a + b$ instead of dividing by $a + b$.
   G. Correct answer.
   H. The student subtracts $2(a + b)$ from the left side of the equation instead of multiplying by 2 and dividing by $a + b$.
   I. The student divides the left side of the equation by 2 instead of multiplying by 2.

6. **A.** The student adds 9 and 20 instead of subtracting.
   B. The student adds 9 and 20 instead of subtracting, and then miscalculates $5x - x$, ignoring the coefficient of x.
   C. Correct answer
   D. The student calculates $20 - 9$ correctly, but then miscalculates $5x - x$, ignoring the coefficient of x.

Chapter 1 Standards Assessment Item Analysis (continued)

7. **F.** The student only accounts for one book on the right side.

   G. Correct answer

   H. The student adds the weight of the coffee can to that of the books instead of subtracting it.

   I. The student correctly subtracts the coffee can from the two books but fails to divide the result by 2.

8. **4 points** The student demonstrates a thorough understanding of setting up and solving equations. In Part A, the student writes the equation $20x = 4900$ and gets $x = 245$. In Part B, the student writes the equation $22y + 650 = 5490$ and gets $y = 220$. In Part C, the student writes the equation $245t = 220t + 650$ and gets $t = 26$.

   3 points The student demonstrates an essential but less than thorough understanding of setting up and solving equations. The equations are set up correctly but there is a small calculation error. If an error is made in Part A or B, the student's work in part C should follow logically from the error.

   2 points The student demonstrates a partial understanding of setting up and solving equations. The student's work and explanations demonstrate a lack of essential understanding. For example, one of the equations is set up wrong.

   1 point The student demonstrates a limited understanding of setting up and solving equations. The student's response is incomplete and exhibits many flaws. For example, all of the equations are set up wrong.

   0 points The student provided no response, a completely incorrect or incomprehensible response, or a response that demonstrates insufficient understanding of setting up and solving equations.

Name_____ Date _____

Chapter 1 Alternative Assessment

1. Linda likes to figure out how number puzzles work. Just this week, she learned about three new puzzles and is trying to work them out. See if you can help her.

   a. Puzzle 1

   Think of two different digits. Add them to get Sum 1.
   Form two different 2-digit numbers with the digits. Add them to get Sum 2.
   Divide Sum 2 by Sum 1.

   Use equations to show why the quotient will always be 11.

   b. Puzzle 2

   I am thinking of a number. Multiplying it by 7 and adding 2 is equal to multiplying it by 2 and adding 7. What is the number?

   Use an equation to find the answer. Then explain or show why you can substitute any number you choose both for 7 and for 2.

   c. Puzzle 3

   Choose a number. Multiply it by 4. Add 8 to the product. Divide the sum by 4. Tell me what number you have now.

   What must be done to figure out what number you chose? Use algebra to show how to find the original number choice.

2. Jake and Cal were bragging about how fast they could run. Jake announced that he ran $2\frac{1}{2}$ miles in half an hour. Cal replied that because he ran 6600 feet in 15 minutes, he ran farther and faster than Jake.

   a. Who ran farther? Explain why.

   b. Who ran faster? Explain why.

Name _____ Date _____

 **Chapter 1 Alternative Assessment Rubric**

Score	Conceptual Understanding	Mathematical Skills	Work Habits
4	Shows complete understanding of: • writing an equation to represent a given situation • solving 1-step and 2-step equations • converting and comparing measures	Writes equations correctly and accurately for the three puzzle situations. Solves the equations correctly for the three puzzles. Correctly converts and compares all the measures in Exercise 2.	Answers all parts of both problems. All equations and expressions are written in a systematic way. Work is very neat and well organized.
3	Shows nearly complete understanding of: • writing an equation to represent a given situation • solving 1-step and 2-step equations • converting and comparing measures	Writes equations correctly and accurately for two puzzle situations. Solves the equations correctly for two puzzles. Correctly converts and compares most of the measures in Exercise 2.	Answers several parts of both problems. Most equations and expressions are written in a systematic way. Work is neat and organized.
2	Shows some understanding of: • writing an equation to represent a given situation • solving 1-step and 2-step equations • converting and comparing measures	Writes an equation correctly and accurately for one puzzle situation. Solves the equation correctly for one puzzle. Correctly converts and compares some measures in Exercise 2.	Answers some parts of both problems. Equations and expressions are written haphazardly. Work is not very neat or organized.
1	Shows little understanding of: • writing an equation to represent a given situation • solving 1-step and 2-step equations • converting and comparing measures	Writes equations correctly and accurately for the three puzzle situations. Does not solve the equation correctly for any puzzle. Does not convert or compare the measures in Exercise 2.	Does not attempt any part of either problem. No equations or expressions are written. Work is sloppy and disorganized.

Name_____ Date _____

Chapter 2 Quiz
For use after Section 2.4

Graph the linear equation using a table.

1. $y = \dfrac{1}{2}x - 3$

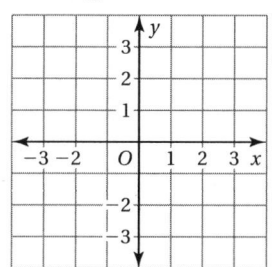

2. $y = -\dfrac{x}{4} + 1$

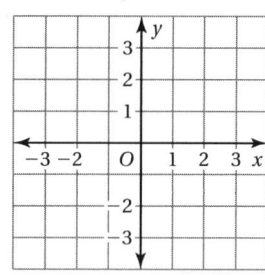

Find the slope of the line.

3.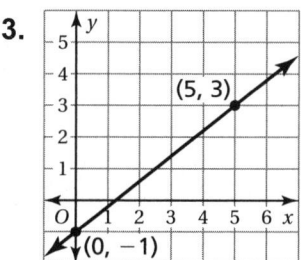

4.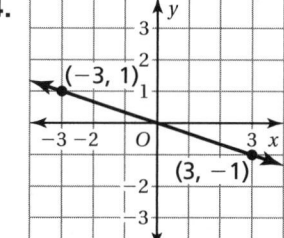

5. What is the slope of a line that is perpendicular to the line in Exercise 3?

Find the slope and y-intercept of the graph of the equation.

6. $y = -4x - 6$

7. $y = \dfrac{1}{2}x - \dfrac{1}{3}$

Find the x- and y-intercepts of the graph of the equation.

8. $3x - 4y = 24$

9. $-6x + 3y = 12$

10. It costs $2.50 to rent a pair of bowling shoes and $1.75 for each game bowled. Write a linear equation that models the cost y of bowling x games. Graph the equation.

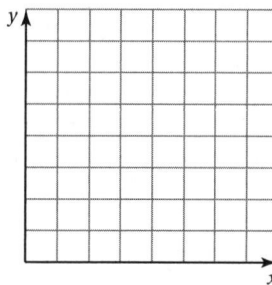

11. You spend $50 on a meal for you and your friends.

   a. Graph the equation $1.5y + 4x = 50$, where x is the number of sandwiches purchased and y is the number of beverages purchased.

   b. Interpret the intercepts.

Answers

1. _____See left._____
2. _____See left._____
3. _____
4. _____
5. _____
6. _____
   _____
7. _____
   _____
8. _____
   _____
9. _____
   _____
10. _____
    _____See left._____
11. a. ___See left.____
    b. _____
    _____
    _____
    _____

Name _____ Date _____

Chapter 2 Quiz
For use after Section 2.7

Write an equation of the line in slope-intercept form.

1.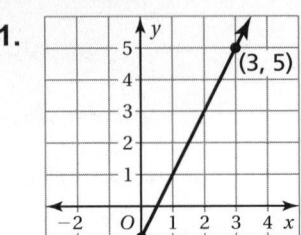

2.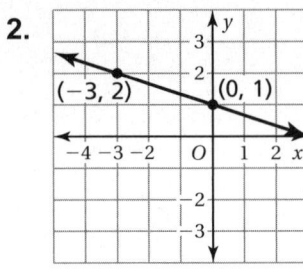

Write an equation of the line with the given slope that passes through the given point.

3. $m = -2; (1, 2)$

4. $m = \dfrac{1}{4}; (4, -2)$

5. $m = 1; (-3, -2)$

6. $m = -\dfrac{1}{3}; (9, -1)$

Write an equation of the line that passes through the points.

7. $(-3, -1), (4, -1)$

8. $(2, 5), (0, 1)$

9. Write an equation of the line that passes through $(4, -3)$ and is (a) parallel to and (b) perpendicular to the line $y = 6x - 1$.

10. A stock is worth $21. Its value is increasing at a rate of $0.25 per week. Write an equation for the value y (in dollars) of the stock after x weeks.

11. You pay $620 to rent a hotel room for 4 weeks. The total cost includes an initial deposit plus a weekly fee of $125.

    a. Write an equation that represents your total cost y (in dollars) after x weeks.

    b. Interpret the y-intercept.

12. You are draining your fish aquarium. After 2 minutes, there are 6 gallons of water in the aquarium. After 5 minutes, the aquarium is empty. Write an equation that represents the volume y (in gallons) of water in the aquarium after x minutes.

Answers

1. _____
2. _____
3. _____
4. _____
5. _____
6. _____
7. _____
8. _____
9. a. _____
   b. _____
10. _____
11. a. _____
    b. _____
    _____
12. _____

Name _____ Date _____

Chapter 2 Test A

Complete the table. Plot the two solution points and draw a line *exactly* through the two points. Find a different solution point on the line. (Use the same axes for both graphs.)

Answers

1. _____See left._____

1.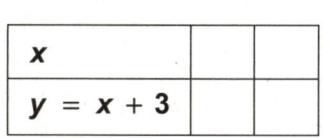

2. _____See left._____

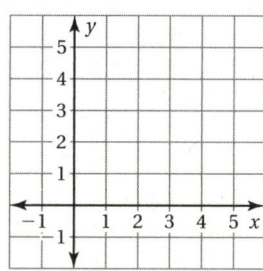

2.

x		
y = x + 3		

3. _____

4. _____

Solve for y.

3. $x + 4y = -12$

4. $2x - 3y = 3$

5. _____

6. _____

Find the slope of the line.

7. _____

5.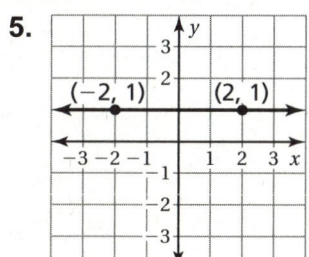

6.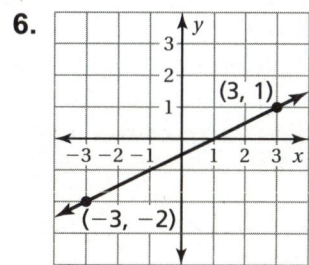

8. _____

9. _____

10. _____

7. Which is steeper, a slide that rises 3 feet for every 2 feet of run, or a sliding pole that rises 5 feet for every 3 feet of run? Explain.

11. _____

8. The equation of a line is $y = 2x - 3$. Write the equation of a line parallel to this line.

12. _____

Find the slope and y-intercept of the graph of the linear equation.

9. $y = 3x - 6$

10. $y + 5 = -\dfrac{3}{4}x$

11. $y = \dfrac{7}{9}x - 3\dfrac{1}{3}$

12. The position y (in meters) of a submarine after x minutes is $y = -8x - 12$. Interpret the y-intercept and the slope.

Chapter 2 Test A (continued)

Graph the linear equation.

13. $-2x + 4y = 12$

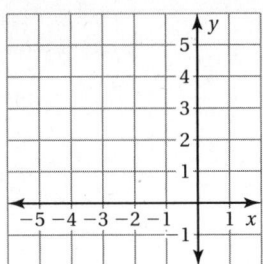

14. $2x + y = -4$

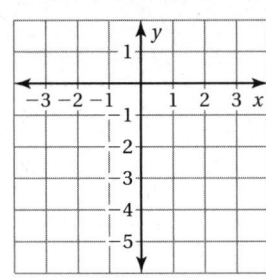

15. You are 9 miles away from home. You start biking home at a speed of 6 miles per hour.

   a. Write an equation in standard form that represents your distance from home y after x hours.

   b. Find the y-intercept of the graph. What does this represent?

   c. Find the x-intercept of the graph. What does this represent?

Write an equation in slope-intercept form of the line that passes through the given points.

16. $(0, 1), (2, 4)$

17. $(-3, 1), (0, 4)$

18. $(-3, 7), (2, -3)$

19. $(2, 8), (-2, 10)$

20. The graph shows the height y (in feet) of a kite x seconds after you start letting out the string.

   a. Find and interpret the slope of the graph.

   b. Write an equation of the line of the graph.

   c. What is the height of the kite after 15 seconds?

   d. Interpret the y-intercept of the graph.

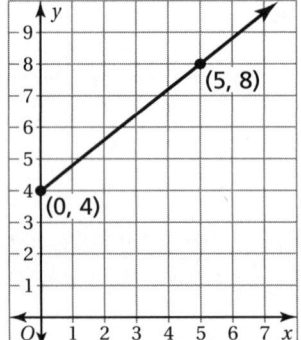

Answers

13. _See left._

14. _See left._

15. a. _____

   b. _____

   c. _____

16. _____

17. _____

18. _____

19. _____

20. a. _See left._

   b. _____

   c. _____

   d. _See left._

Chapter 2 Test B

Solve for y. Then graph the equation.

1. $3x + 2y = -4$

2. $4y - 3x = 4$

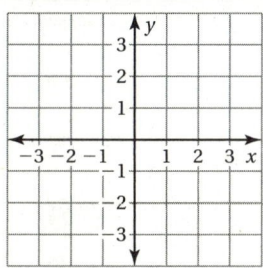
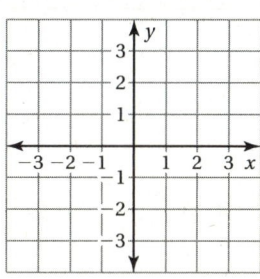

Find the slope of the line.

3.

4.

5. Which is steeper, a hill that rises 2 feet for every 10 feet of run, or a hill that rises 2 feet for every 15 feet of run? Explain.

6. Which two lines are parallel? Explain.

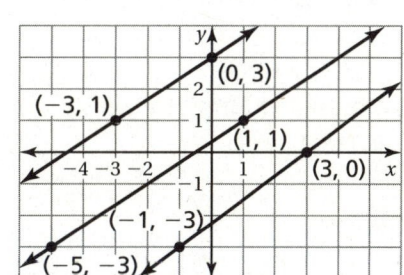

Find the slope and y-intercept of the graph of the linear equation.

7. $y = -2x - 1$

8. $y - \dfrac{1}{3}x = 0$

9. $y + 2 = \dfrac{3}{4}x$

10. Explain how to find the x-intercept of the graph of $y = 4x - 2$.

Answers

1. _____
   See left.
2. _____
   See left.
3. _____
4. _____
5. _____
6. _____
7. _____
8. _____
9. _____
10. _____

Chapter 2 Test B (continued)

Find the *x*-intercept and the *y*-intercept. Graph the equation.

11. $3x - 2y = 6$

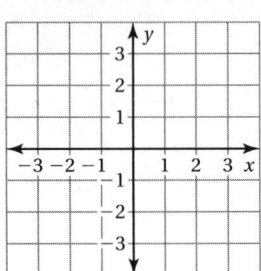

12. $2x - y = -2$

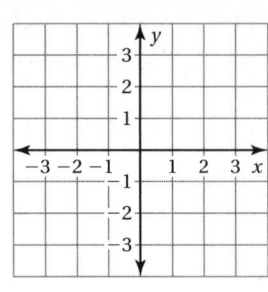

13. You borrow $90 from your grandmother. You pay back $15 each week.

   a. Write an equation in standard form that represents the amount owed *y* after *x* weeks.

   b. Find the *y*-intercept of the graph. What does this represent?

   c. Find the *x*-intercept of the graph. What does this represent?

Write an equation of the line that passes through the points.

14. $(-3, -2), (0, 0)$

15. $(0, 3), (2, 3)$

16. $(-4, -3), (-2, 2)$

17. $(9, -5), (6, 4)$

18. The graph shows the relationship between temperature *y* (in degrees Fahrenheit) and altitude *x* (in thousands feet).

   a. Find and interpret the slope of the graph.

   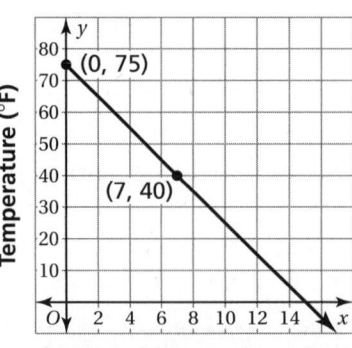

   b. Write an equation of the line.

   c. Interpret the *x*-intercept of the graph.

   d. What is the temperature at 11,000 feet?

Answers

11. _____

   _____

   See left.

12. _____

   _____

   See left.

13. a. _____

   b. _____

   _____

   _____

   c. _____

   _____

14. _____

15. _____

16. _____

17. _____

18. a. **See left.**

   b. _____

   c. **See left.**

   d. _____

Name_____ Date_____

Chapter 2 Standards Assessment

1. Carla plotted the points on the graph below to show how the amount she owes for tuition decreases as the number of tuition payments increases. The slope of the line segment joining these points is $-\frac{2}{3}$.

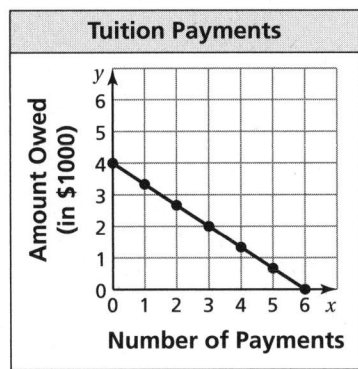

   What does the slope of the line segment represent?

   A. Each payment decreases the amount owed by $4,000.

   B. Each payment decreases the amount owed by $0.66.

   C. For every 3 payments, the amount owed decreases by $2,000.

   D. For every 2 payments, the amount owed decreases by $3,000.

2. **GRIDDED RESPONSE** What value of k makes the equation below true?

   $$5k - 12 = 22$$

3. A line contains the points $(0, 9)$ and $(6, 6)$. Which point is also on this line?

   F. $(2, 5)$ H. $(-8, 5)$

   G. $(4, 7)$ I. $(4, 4)$

4. Chris borrowed money from her brother. Each week she pays him $5 toward her debt. After 8 weeks, she has $8 left to pay. This situation is modeled by a line with a slope of -5 that contains the point $(9, 3)$. At what point does this line pass through the y-axis?

   A. $(0, 45)$ C. $(0, -42)$

   B. $(0, 24)$ D. $(0, 48)$

Chapter 2 Standards Assessment (continued)

5. Vivian charges $4 for bracelets and $5 for earrings. Her cost to make x bracelets and y earrings is $60. The equation $4x + 5y = 60$ represents this situation. The graph of this equation is a line. What is the slope of the line?

   F. -4 **H.** 0.8

   G. -0.8 **I.** 12

6. The math teacher asked Edith, "How old are you?" "Sixty years less than five times my brother's age," she answered. "That doesn't help me," replied the teacher. "Yes, it does," said Edith, "He and I are twins!" How old is Edith?

   A. 10 **C.** 15

   B. 12 **D.** 55

7. A line passes through the point $(1, 3)$ and has a slope of 2. Which of these points also lies on this line?

   F. $(1, 5)$ **H.** $(3, 5)$

   G. $(2, 6)$ **I.** $(3, 7)$

8. **SHORT RESPONSE** A car is traveling at a speed of 45 miles per hour. Once the car starts to brake, its speed (s) is related to the number of seconds (t) it spends braking according to the formula shown below.

   $$s = -10t + 45$$

   Part A Draw and label a graph that represents this situation on the coordinate grid below.

   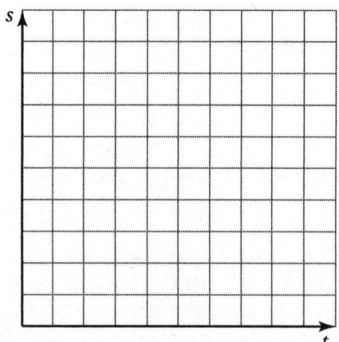

   Part B Determine how many seconds it takes for the car to stop.

   Time to stop _____ seconds

Chapter 2 Standards Assessment Item Analysis

1. **A.** The student mistakes the y-intercept for the slope.
   B. The student realizes the slope is $-\frac{2}{3}$, but doesn't realize the y-axis is in 1,000s, so each payment decreases the amount owed by $666.67.
   C. Correct answer
   D. The student confuses the rise and run in the slope.

2. Correct answer: 6.8
   Common error: The student subtracts 12 from the right hand side instead of adding, leading to an incorrect answer of $k = 2$.

3. **F.** The student miscalculates the slope to be -2 and gets the equation $y = -2x + 9$.
   G. Correct answer
   H. The student miscalculates the slope to be $\frac{1}{2}$ and gets the equation $y = \frac{1}{2}x + 9$.
   I. The student correctly calculates the slope to be $-\frac{1}{2}$, but uses 6 for the y-intercept and gets the equation $y = -\frac{1}{2}x + 6$.

4. **A.** The student lets $y_1 = 0$ in the equation $y - y_1 = m(x - x_1)$.
   B. The student switches x_1 and y_1 in the equation $y - y_1 = m(x - x_1)$.
   C. The student lets $m = 5$ in the equation $y - y_1 = m(x - x_1)$.
   D. Correct answer

5. **F.** The student correctly subtracts $4x$ from each side but then fails to divide each side by 5.
   G. Correct answer
   H. The student divides 4 by 5 correctly but misplaces the minus sign, possibly by moving $4x$ to the right side incorrectly.
   I. The student divides 60 by 5, getting the y-intercept instead of the slope.

6. **A.** The student writes and solves the incorrect equation $60 - 5x = x$.
   B. The student writes the correct equation $5x - 60 = x$, but makes a mistake subtracting $5x$ and x, yielding $5x$ (because of the hidden coefficient on x) instead of $4x$.
   C. Correct answer
   D. The student misunderstands the problem and simply finds the difference between 60 and 5.

Chapter 2 Standards Assessment Item Analysis (continued)

7. **F.** The student adds 2 to the *y*-coordinate to get another point because the slope is 2.

   G. The student multiplies both coordinates by 2 to get another point because the slope is 2.

   H. The student adds 2 to both coordinates to get another point because the slope is 2.

   I. Correct answer

8. **2 points** The student demonstrates a thorough understanding of translating a situation into a graph and then using the graph (or its related equation) to find the coordinates of a point on that graph. The graph is clearly labeled, properly scaled, and accurately drawn, showing a line from (0, 45) to (4.5, 0). The car stops after 4.5 seconds, and the student indicates this.

   1 point The student's work demonstrates limited understanding of translating a situation into a graph and then using the graph (or its related equation) to find the coordinates of a point on that graph. Either the graph is drawn incorrectly or the student cannot identify the correct point in Part B from a properly drawn graph.

   0 points The student provides no response, a completely incorrect or incomprehensible response, or a response that demonstrates insufficient understanding of making and interpreting graphs.

Chapter 2 Alternative Assessment

1. Sophia is going to make a tomato and fresh mozzarella cheese salad for a party. When she went to the farmers market, she saw that the tomatoes she wanted were $2.50 per pound and that the fresh mozzarella cheese was $5.00 per pound. She is wondering what combinations of whole pounds of tomatoes and mozzarella cheese she can buy with $20.

   a. Write an equation that represents the pounds t of tomatoes and pounds m of mozzarella cheese she can buy with $20.

   b. Graph the equation. Interpret the intercepts.

   c. What combinations of whole pounds of the two ingredients can Sophia buy?

   d. Her recipe uses half as much mozzarella cheese as tomatoes. How much of each ingredient did she buy?

2. The graph shows the height of a Willow Oak Tree.

   a. Find the slope of the line.

   b. Explain the meaning of the slope as a rate of change.

   c. Write an equation of the line.

   d. What is the height of the tree after 30 years?

   e. If the tree growth is constant, when will the tree reach its maximum height of 60 feet?

   f. How would the graph change if the tree grew at a rate of 2 feet per year?

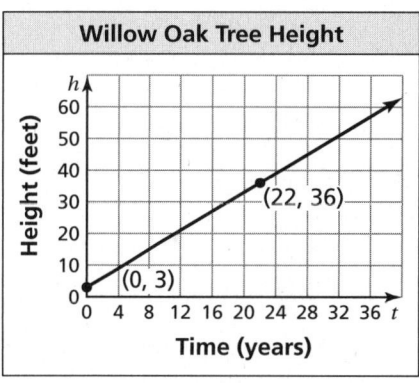

Name _____ Date _____

Chapter 2 Alternative Assessment Rubric

Score	Conceptual Understanding	Mathematical Skills	Work Habits
4	Shows complete understanding of: • identifying slope, x-intercept, and y-intercept of equations graphically and algebraically • using an equation to represent a real-world situation	Writes and graphs the equation and interprets the graph to answer all of Exercise 1. Find the slope, writes the equation, and interprets the graph to answer all of Exercise 2.	Answers all parts of both problems. All equations and graph are written or drawn carefully and systematically. Work is very neat and well organized.
3	Shows nearly complete understanding of: • identifying slope, x-intercept, and y-intercept of equations graphically and algebraically • using an equation to represent a real-world situation	Writes and graphs the equation and interprets the graph to answer most of Exercise 1. Find the slope, writes the equation, and interprets the graph to answer most of Exercise 2.	Answers several parts of both problems. Most equations and graph are written or drawn carefully and systematically. Work is neat and organized.
2	Shows nearly complete understanding of: • identifying slope, x-intercept, and y-intercept of equations graphically and algebraically • using an equation to represent a real-world situation	Writes and graphs the equation and interprets the graph to answer some of Exercise 1. Find the slope, writes the equation, and interprets the graph to answer some of Exercise 2.	Answers some parts of both problems. Equations and graph are written or drawn carelessly. Work is not very neat or organized.
1	Shows little understanding of: • identifying slope, x-intercept, and y-intercept of equations graphically and algebraically • using an equation to represent a real-world situation	Does not answer Exercise 1. Does not solve the system of linear equations either graphically or algebraically. Identifies no features of either equation and does not rewrite either in standard form.	Does not attempt any part of either problem. No equations or graphs are written or drawn. Work is sloppy and disorganized.

Name_____ Date_____

Chapter 3 Quiz
For use after Section 3.3

Write the word sentence as an inequality.

1. A number c minus 12 is greater than 4.

2. A number y plus 3.6 is no more than 9.5.

Tell whether the given value is a solution of the inequality.

3. $x - 2 \geq 6$; $x = 8$
4. $3c < 36$; $c = 13$

Graph the inequality on a number line.

5. $x \geq -2$
6. $a > 1.5$
7. $k < \dfrac{2}{3}$

Solve the inequality. Graph the solution.

8. $y + 4 \leq 7$
9. $-15z \leq -45$
10. $\dfrac{d}{-2} > 4$

Write the word sentence as an inequality. Then solve the inequality.

11. The product of a number and -5 is at least 35.

12. A number divided by 3 is no more than 12.

13. A person that is at least 65 years old is often considered a senior citizen. Write an inequality that represents this situation.

14. The solution of $x + b > -14$ is $x > -21$. What is the value of b?

15. Your gas tank can hold no more than 14.5 gallons of gasoline. On a trip to the grocery store, you use 1.5 gallons of gasoline. Write and solve an inequality that represents the amount of gasoline left in your gas tank.

16. You need to score at least 1500 points on your new video game to obtain the high score. You get 300 points after completing each level. Write and solve an inequality to find the number of levels you must beat in order to obtain the high score.

Answers

1. _____
2. _____
3. _____
4. _____
5. _See left._
6. _See left._
7. _See left._
8. _____
   See left.
9. _____
   See left.
10. _____
    See left.
11. _____
    _____
12. _____
    _____
13. _____
14. _____
15. _____
16. _____
    _____

Name _____ Date _____

Chapter 3 Quiz
For use after Section 3.5

Solve the inequality. Graph the solution.

1. $3t - 1 < 8$

2. $1.6w + 1.7 \geq 4.9$

Write the word sentence as an inequality. Graph the inequality.

3. A number j is greater than 2 and less than or equal to 5.

4. A number n is greater than or equal to 4 or at most -2.

Solve the inequality. Graph the solution, if possible.

5. $4c < 36$ or $5c \geq 60$

6. $|2k + 3| > 1$

Graph the inequality in a coordinate plane.

7. $x < 3$

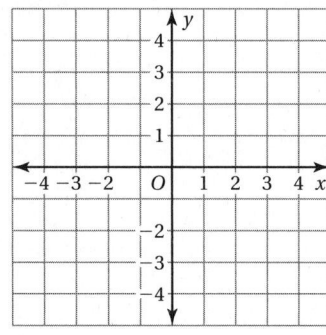

8. $y \leq -2x - 1$

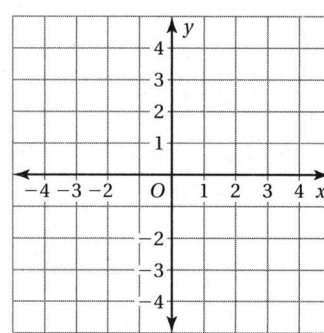

9. At an arcade, small prizes cost 50 tickets each. You have 350 tickets. Write and solve an inequality that represents the number of small prizes you can purchase with your tickets and still have 100 tickets left over.

10. At a carnival, the dart game costs $0.50 to play and the ring toss game costs $0.25 to play. You have $5 to spend on the two games. Write an inequality for the number of times you can play the two games.

Answers

1. _____
   See left.

2. _____
   See left.

3. _____
   See left.

4. _____
   See left.

5. _____
   See left.

6. _____
   See left.

7. _____ See left.

8. _____ See left.

9. _____
   _____

10. _____

Name_____ Date_____

 **Chapter 3 Test A**

Write an inequality for the graph.

Answers

1. ←——○——+——+——+——→
   −2 −1   0   1   2

2. ←——+——+——●——+——→
   −4 −3 −2 −1   0

1. _____

2. _____

Write the word sentence as an inequality.

3. A number *n* is no less than −3.

4. A number *q* plus 7 is less than 45.

5. A number *x* divided by −1 is at least −4.

6. The children in the class are more than 10 years old.

7. The minimum cost for parking is $3.

3. _____
4. _____
5. _____
6. _____
7. _____
8. _____

Tell whether the given value is a solution of the inequality.

8. $j + 1 > 10$; $j = 9$

9. $-3 \leq \dfrac{k}{2}$; $k = -1$

9. _____
10. ___See left.___
11. a._____

10. A freezer is set to turn on and start cooling if the temperature rises above −10° Celsius. The cooling turns off when the freezer has reached a temperature of −16° Celsius. Write two inequalities to model the situation. Give a sample value at which the cooling would turn on, and a sample value at which the cooling would be off.

b._____

c._____

11. An elevator can carry 800 pounds of weight.

  a. A student weighing 95 pounds gets on the elevator. Write and solve an inequality to represent the remaining weight that can be added.

  b. A football player weighing 280 pounds gets on the elevator with the student. Write and solve an inequality representing the remaining weight that can be added.

  c. Two more football players weighing a total of 470 pounds come to the elevator. Can they get on safely? Explain.

Name _____ Date _____

Chapter 3 Test A (continued)

Solve the inequality.

12. $x - 3 > 7$ 13. $m + 2 \leq -4$ 14. $6y > 8$

15. $p \div 5 < -3$ 16. $4z - 3 \geq -1$ 17. $3 < t + 2 < 6$

Solve the inequality. Graph the solution.

18. $-4 + x \leq 1$

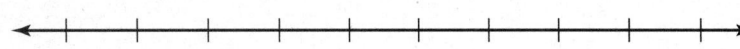

19. $2 < -\dfrac{y}{5}$

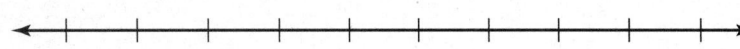

20. $3|4x + 2| \geq 6$

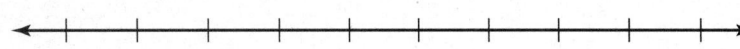

Graph the inequality in the coordinate plane.

21. $x < 2$ 22. $4x + 2y \leq 4$

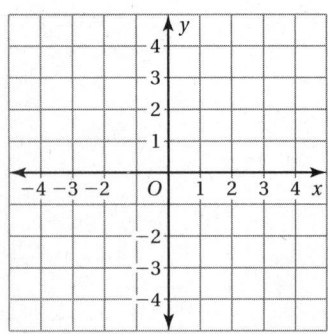   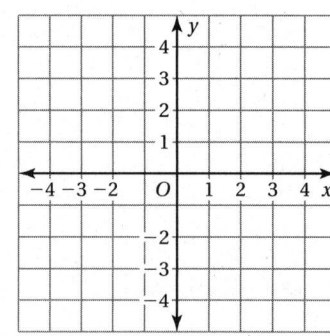

23. The basketball team spends 20 minutes running laps and at least 15 minutes discussing plays. Practice lasts one hour and 45 minutes. Write an inequality to represent the amount of time to work on other drills.

Answers

12. _____
13. _____
14. _____
15. _____
16. _____
17. _____
18. _____
     See left.
19. _____
     See left.
20. _____
     See left.
21. _See left._
22. _See left._
23. _____
     _____

Name _____ Date _____

Chapter 3 Test B

Write the word sentence as an inequality.

1. A number x is less than $\frac{1}{4}$.
2. A number n is no more than 8.
3. A number m minus 3 is more than -4.
4. Sixteen times a number j is no less than -2.
5. Twice a number q minus 1 is less than 5.
6. A number a divided by 2 is no more than 6.
7. To pass the test you must score at least 60 on the test.
8. The maximum cost is $35.

Tell whether the given value is a solution of the inequality.

9. $\frac{x}{2} - 1 < -1$; $x = -\frac{3}{4}$
10. $5x - 17 > 62$; $x = 13$

11. A video game gives you 100 seconds to complete the level and move to the next. You are halfway through the level after 55 seconds.

   a. Write and solve an inequality to find out how much time you have left to complete the level.

   b. You will receive a time bonus if you finish in 70 seconds or less. Write and solve an inequality to find how much time you have left to earn a time bonus.

   c. You finish the game in another 32 seconds. Do you earn a time bonus? Do you move to the next level? Explain.

12. An isosceles triangle has a base of 5 centimeters and legs x centimeters long. The perimeter is no more than 30 centimeters. Write and solve an inequality to find the possible values of x.

Solve the inequality.

13. $b + 8 > 7$
14. $3 \geq x + 4.5$
15. $7c \leq 35$
16. $\frac{p}{3} > -5$
17. $\frac{1}{4}(w - 5) \geq -2$
18. $6|2n + 4| < 24$

Answers

1. _____
2. _____
3. _____
4. _____
5. _____
6. _____
7. _____
8. _____
9. _____
10. _____
11. a. _____
    _____
    _____
    _____
    b. _____
    _____
    _____
    c. _See left._
12. _____
    _____
13. _____
14. _____
15. _____
16. _____
17. _____
18. _____

Chapter 3 Test B (continued)

Solve the inequality. Graph the solution.

19. $w - 8 \geq -4$

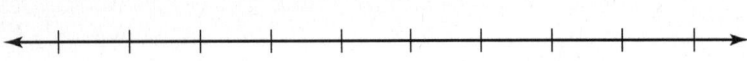

20. $-4 > -\dfrac{m}{10}$

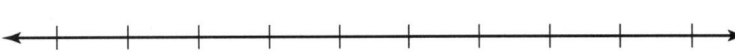

21. $\dfrac{z-1}{2} \geq \dfrac{1}{3}$

22. $4(x - 3) < -20$ or $2(3x - 2) > -10$

Graph the inequality in the coordinate plane.

23. $y \geq -3$

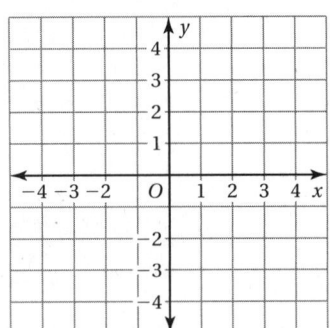

24. $14x - 7y > 21$

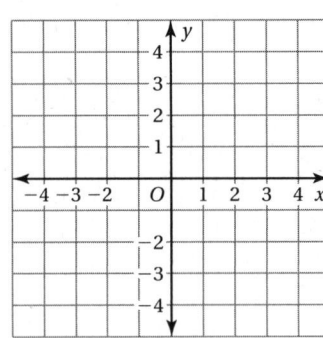

25. A music teacher budgets $150 for new books. The minimum cost of a new book is $12. How many books can she buy? Is this a minimum or a maximum amount? Explain.

Answers

19. _____
    See left.

20. _____
    See left.

21. _____
    See left.

22. _____
    See left.

23. See left.

24. See left.

25. _____

Chapter 3 Standards Assessment

1. Which inequality is represented by the graph shown below?

   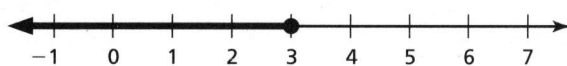

   A. $x > 3$ C. $x < 3$

   B. $x \geq 3$ D. $x \leq 3$

2. A linear equation relating temperature in degrees Fahrenheit y to temperature in degrees Celsius x is $y - 32 = \frac{9}{5}x$. What are the slope and y-intercept of the graph of this equation?

   F. The slope is $\frac{9}{5}$ and the y-intercept is 32.

   G. The slope is $\frac{9}{5}$ and the y-intercept is -32.

   H. The slope is 32 and the y-intercept is $\frac{9}{5}$.

   I. The slope is -32 and the y-intercept is $\frac{9}{5}$.

3. **GRIDDED RESPONSE** The triangle and rectangle shown below have the same perimeter. What is the area of the rectangle in square feet?

   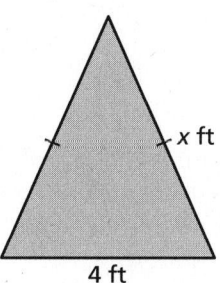
   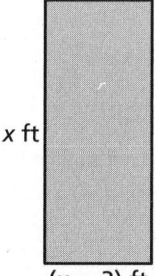

4. A skyscraper contains 80 floors, with adjacent floors 10 feet apart. An elevator in the skyscraper is ascending at a rate of 5 feet per second. How long will it take the elevator to ascend from the 4th floor to the 28th floor?

   A. 12 seconds   C. 48 seconds

   B. 14 seconds   D. 56 seconds

Name _____ Date _____

Chapter 3 Standards Assessment (continued)

5. Which inequality represents the following sentence?

   | A number d is no less than -4 and fewer than 7. |

   F. $-4 < d < 7$ H. $-4 \le d < 7$

   G. $-4 < d \le 7$ I. $d \ge -4$ or $d < 7$

6. Which equation passes through the point $(1, 1)$ and is perpendicular to the line shown in the graph?

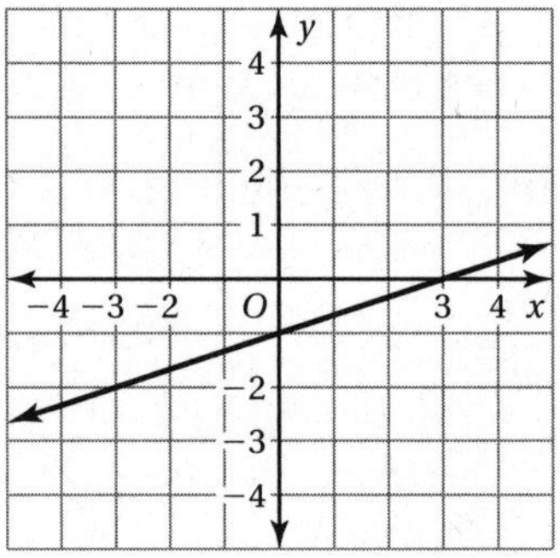

   A. $y = -3x$

   C. $y = -3x + 4$

   B. $y = -\dfrac{1}{3}x + \dfrac{4}{3}$

   D. $y = 3x - 2$

7. What is the solution for the inequality shown below?

   $$-4.5 > 3 + x$$

   F. $x < -7.5$ H. $x > -7.5$

   G. $x < -1.5$ I. $x > -1.5$

8. **EXTENDED RESPONSE** A freight elevator can hold a maximum weight of 3,500 pounds.

   Part A Let w be the weight allowed on the elevator. Write an inequality that represents this situation.

   Inequality _____

   Part B A deliveryman weighs 200 pounds. He is delivering cartons that each weigh 48 pounds. He wants to know how many cartons he can safely put on the elevator at one time. Let c represent the number of cartons. Write an inequality that represents this situation.

   Inequality _____

   Part C Solve your inequality from Part B. Show your work fully. Explain what your solution means.

Chapter 3 Standards Assessment Item Analysis

1. **A.** The student confuses the meaning of the symbols and excludes the number 3.
    B. The student confuses the meaning of the symbols.
    C. The student forgets to include 3.
    D. Correct answer

2. **F.** Correct answer
    G. The student subtracts 32 when solving for y instead of adding.
    H. The student switches slope with y-intercept.
    I. The student subtracts 32 when solving for y instead of adding and switches slope with y-intercept

3. Correct answer: 10

    Common error: The student correctly computes $x = 5$ but finds the perimeter, 14, instead of the area.

4. **A.** The student inverts the relationship between 5 feet per second and 10 feet per floor.
    B. The student uses 28 floors instead of 24 floors as the distance and inverts the relationship between 5 feet per second and 10 feet per floor.
    C. Correct answer.
    D. The student calculates the time to travel 28 floors.

5. **F.** The student excludes the number -4.
    G. The student excludes the number -4 and includes the number 7.
    H. Correct answer
    I. The student forgets that the inequality must include only numbers that satisfy both conditions in the sentence, not just either condition.

6. **A.** The student calculates the correct slope, but ignores the point the line must pass through.
    B. The student incorrectly uses $-\frac{1}{3}$ for the slope.
    C. Correct answer
    D. The student incorrectly uses 3 for the slope.

7. **F.** Correct answer
    G. The student subtracts 3 from -4.5 incorrectly.
    H. The student subtracts 3 from -4.5 correctly but misapplies the rule for multiplying and dividing an inequality by a negative number.
    I. The student subtracts 3 from -4.5 incorrectly and misapplies the rule for multiplying and dividing an inequality by a negative number.

Chapter 3 Standards Assessment Item Analysis
(continued)

8. **4 points** The student demonstrates a thorough understanding of writing and solving inequalities. In Part A, the inequality $w \leq 3{,}500$ is written. In Part B, the inequality $48c + 200 \leq 3{,}500$, or its equivalent, is written. In Part C, the solution $c \leq 68.75$ is obtained and the student points out that the deliveryman can safely load 68 cartons at one time.

   3 points The student demonstrates an essential but less than complete understanding. Either a small error is made in Part B (with consistent work being carried forward in Part C), or the student fails to correctly interpret the solution of 68.75 in Part C.

   2 points The student's work and explanations demonstrate a lack of essential understanding of writing and solving inequalities. Part A is solved correctly but each subsequent part shows errors.

   1 point The student demonstrates limited understanding of writing and solving inequalities. The student's response is incomplete and exhibits many flaws.

   0 points The student provides no response, a completely incorrect or incomprehensible response, or a response that demonstrates insufficient understanding of writing and solving inequalities.

Chapter 3 Alternative Assessment

1. A number z added to 5 is less than or equal to 8.
   A number q subtracted from 5 is greater than or equal to 8.

   a. Write an inequality for each statement.

   b. Solve each inequality you wrote in part (a).

   c. Graph each solution you found in part (b). Explain why the circle you used is closed or open.

   d. What do you notice about the solutions?

   e. Write a word problem that can be answered using one of the inequalities.

2. A number t multiplied by 2 is greater than 4.
   A number v multiplied by -2 is less than 4.

   a. Write an inequality for each statement.

   b. Solve each inequality you wrote in part (a).

   c. Graph each solution you found in part (b). Explain why the circle you used is closed or open.

   d. What do you notice about the solutions?

3. Three times a number m minus 1 is greater than or equal to 2.
   The sum of a number w and 2 is less than or equal to 3.

   a. Write an inequality for each statement.

   b. Solve each inequality you wrote in part (a).

   c. Graph each solution you found in part (b). Explain why the circle you used is closed or open.

   d. What do you notice about the solutions?

Name _____ Date _____

Alternative Assessment Rubric

Score	Conceptual Understanding	Mathematical Skills	Work Habits
4	Shows complete understanding of: • solving and graphing inequalities • describing a real-world situation for an inequality	Writes, solves, and graphs all pairs of inequalities correctly. Described relationships in the solutions of all pairs of inequalities. Wrote a word problem that could be solved using one of the given inequalities.	Answers all parts of all questions. All calculations are done carefully. All work is neat and well organized.
3	Shows nearly complete understanding of: • solving and graphing inequalities • describing a real-world situation for an inequality	Writes, solves, and graphs two pairs of inequalities correctly. Described relationships in the solutions of two pairs of inequalities. Wrote a word problem that could be solved using one of the given inequalities.	Answers almost all parts of all questions. Most of the calculations are done carefully. Most of the work is neat and well organized.
2	Shows some understanding of: • solving and graphing inequalities • describing a real-world situation for an inequality	Writes, solves, and graphs one pair of inequalities correctly. Described relationships in the solutions of one pair of inequalities. Wrote a word problem that could be solved using an inequality, but not one of the given inequalities.	Answers some parts of the three questions. Some calculations are done carefully. Some work is neat and well organized.
1	Shows little understanding of: • solving and graphing inequalities • describing a real-world situation for an inequality	Did not write, solve, or graph any inequality correctly. Did not describe relationships in the solutions of any pair of inequalities. Wrote a word problem that could not be solved using any inequality.	Answers only a few parts of some questions. No calculations are done carefully. All work is sloppy and disorganized.

Name_____ Date_____

Chapter 4 Quiz
For use after Section 4.2

Match the system of linear equations with the corresponding graph. Use the graph to estimate the solution. Check your solution.

Answers

1. $y = x + 1$
   $y = 2x - 1$

2. $y = -x - 1$
   $y = -2x + 1$

A.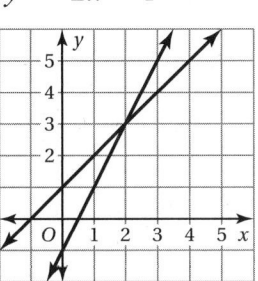

B.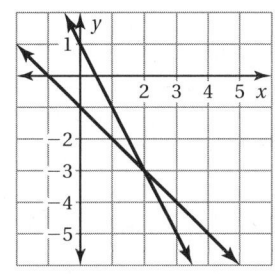

1. _____

_____

2. _____

_____

3. ____See left.____

_____

4. ____See left.____

_____

Solve the system of linear equations by graphing.

3. $y = -x - 2$
   $y = \dfrac{1}{4}x + 3$

4. $-x + 2y = 2$
   $-3x + 4y = 8$

5. _____

6. _____

7. _____

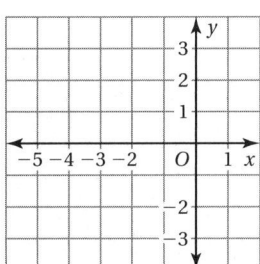

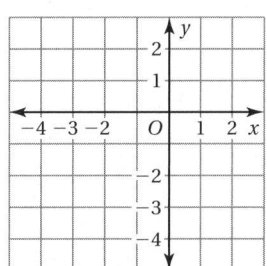

8. _____

_____

9. _____

_____

Solve the system of linear equations by substitution. Check your solution.

5. $y = 2x + 3$
   $y = 3x + 5$

6. $2x + y = 4$
   $-3x + y = -1$

7. $y = \dfrac{1}{3}x + 2$
   $y = \dfrac{1}{6}x + 4$

8. A math club has 40 members. The number of girls is 5 less than 2 times the number of boys. How many members are boys? How many members are girls?

9. The perimeter of a rectangle is 36 feet. The length is 2 less than 3 times the width. What is the length? What is the width?

Name _____ Date _____

Chapter 4 Quiz
For use after Section 4.5

Solve the system of linear equations by elimination. Check your solution.

1. $x + 4y = 4$
   $-x + 2y = 8$

2. $y = -9x + 2$
   $y = -3x - 4$

3. $x + 6y = 12$
   $x + 3y = 3$

Solve the system of linear equations. Check your solution.

4. $-6x + 3y = 9$
   $-8x + 4y = 12$

5. $2x - 3y = 1$
   $-4x + 6y = -4$

6. $-10x + 5y = 30$
   $-2x + 2y = 6$

Use a graph to solve the equation. Check your solution.

7. $-2x + 3 = 4x - 3$

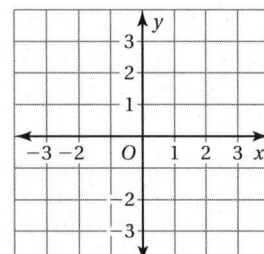

8. $-3x + 3 = -2x + 1$

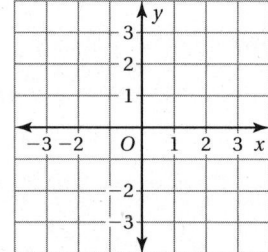

9. Graph the system of linear inequalities.
   $y \geq 2x - 4$
   $y < 3x - 5$

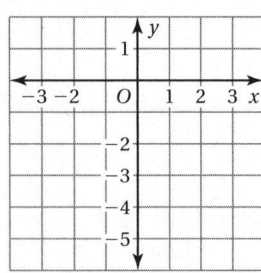

10. Write a system of linear inequalities represented by the graph.

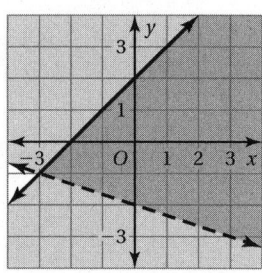

11. A candle shop sells scented candles for $16 each and unscented candles for $10 each. The shop sells 28 candles today and makes $400. How many scented candles did the shop sell today? How many unscented candles did the shop sell today?

12. You have two tests tomorrow. You plan to study at least 4 hours tonight. You want to study Social Studies for more than half the time you study English plus 1 hour. Write a system of linear inequalities that represents this situation.

Answers

1. _____
2. _____
3. _____
4. _____
5. _____
6. _____
7. ___See left.___
   _____
8. ___See left.___
   _____
9. ___See left.___
10. _____
    _____
11. _____
    _____
12. _____
    _____

Name_____ Date_____

Chapter 4 Test A

Match the system of linear equations with the corresponding graph. Use the graph to estimate the solution. Check your solution.

1. $y = 3x - 2$
   $y = 4x - 3$

2. $y = -2x - 3$
   $y = 3x + 2$

A.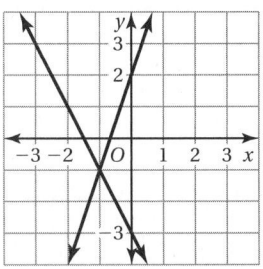

B.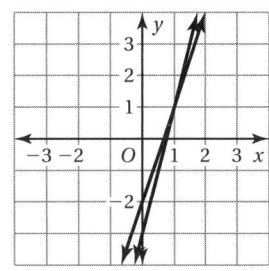

Solve the system of linear equations by graphing.

3. $y = -2x + 1$
   $y = 2x - 3$

4. $y = 6x - 3$
   $y = 4x - 1$

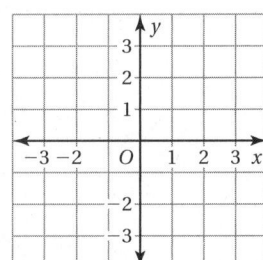

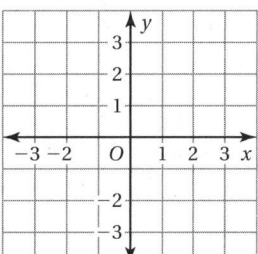

Solve the system of linear equations by substitution. Check your solution.

5. $y = x + 2$
   $y = 3x - 4$

6. $y = 3x + 4$
   $x + y = 8$

7. $-2x + 3y = 9$
   $y = 2x + 7$

8. There are 27 red or blue marbles in a bag. The number of red marbles is 5 less than 3 times the number of blue marbles. How many red marbles are in the bag? How many blue marbles are in the bag?

9. A fruit vendor sells 60 pieces of fruit that are either apples or oranges. The ratio of apples to oranges is 3 : 2. How many apples did the vendor sell? How many oranges did the vendor sell?

Answers

1. _____
   _____
2. _____
   _____
3. ___See left.___
   _____
4. ___See left.___
   _____
5. _____
6. _____
7. _____
8. _____
   _____
9. _____
   _____

Name _____ Date _____

Chapter 4 Test A (continued)

Solve the system of linear equations by elimination. Check your solution.

10. $y = 5x - 8$
    $y = -6x + 3$

11. $2x + 10y = -20$
    $-x + 4y = 28$

12. $-x + 5y = 20$
    $12y = 2x + 60$

Without graphing, determine whether the system of linear equations has *one solution*, *infinitely many solutions*, or *no solution*. Explain your reasoning.

13. $y = 4x + 6$
    $2y = 8x + 12$

14. $y = 3x + 5$
    $y = 3x - 5$

15. $y = 2x + 7$
    $y = 3x - 1$

Use a graph to solve the equation. Check your solution.

16. $2x + 3 = 4x + 5$

17. $-3x - 3 = 2x + 2$

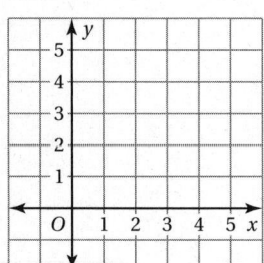
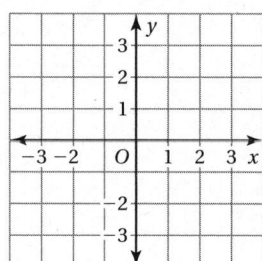

Graph the system of linear inequalities.

18. $y \leq 2x - 3$
    $y \geq 3x - 4$

19. $y < -4x + 2$
    $y < 2x + 2$

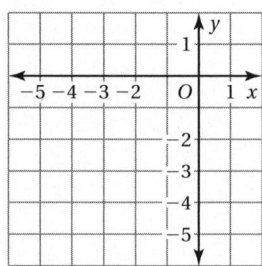
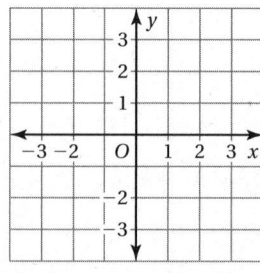

20. At a restaurant, the price for 3 salads and 2 glasses of lemonade is $14 and the price for 2 salads and 4 glasses of lemonade is $12. How much does it cost for 1 salad and 2 glasses of lemonade?

Answers

10. _____
11. _____
12. _____
13. _____
    _____
    _____
14. _____
    _____
    _____
15. _____
    _____
    _____
16. ___See left.___
    _____
17. ___See left.___
    _____
18. ___See left.___
19. ___See left.___
20. _____

Name_____ Date_____

Chapter 4 Test B

Solve the system of linear equations by graphing.

1. $y = x + 3$
   $y = -x - 3$

2. $y = \dfrac{1}{3}x + 2$
   $y = 2x - 3$

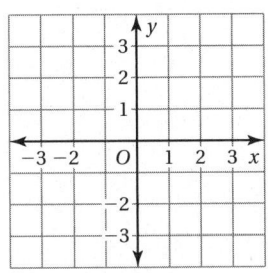
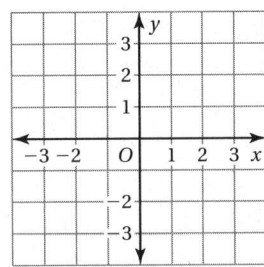

3. $y = 6x - 5$
   $y = 5x - 4$

4. $x + 3y = 6$
   $4x - 6y = 6$

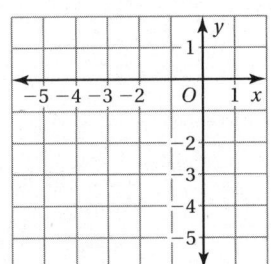
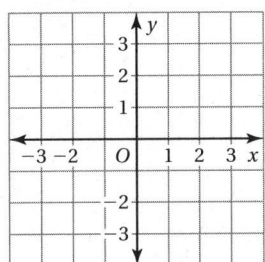

Solve the system of linear equations by substitution. Check your solution.

5. $y = x + 3$
   $y = 4x - 6$

6. $x + y = 7$
   $y = 2x + 4$

7. $x - 3y = -12$
   $y = 2x + 9$

8. There are 24 pens in a desk drawer. The pens are either red or blue. The ratio of red pens to blue pens is 5 : 1. How many pens are red? How many pens are blue?

9. There are 31 fish in a tank. The fish are either orange or red. There are 7 more orange fish than half the number of red fish. How many fish are orange? How many fish are red?

10. The measure of $\angle 1$ is 30 degrees less than twice the measure of $\angle 2$. What is the measure of $\angle 1$? What is the measure of $\angle 2$?

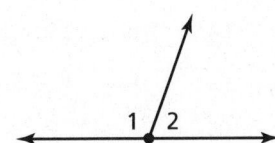

Answers

1. _____See left._____

2. _____See left._____

3. _____See left._____

4. _____See left._____

5. _____

6. _____

7. _____

8. _____

9. _____

10. _____

Name _____ Date _____

Chapter 4 Test B (continued)

Solve the system of linear equations by elimination. Check your solution.

11. $3x - y = 2$
    $2x - y = 3$

12. $2x - y = -2$
    $x - 2y = -16$

13. $6x - 2y = 10$
    $10x - y = -2$

Without graphing, determine whether the system of linear equations has *one solution*, *infinitely many solutions*, or *no solution*. Explain your reasoning.

14. $y = 11x + 8$
    $y = 9x - 7$

15. $14y = x + 6$
    $28y = 2x + 12$

16. $y = 15x + 1$
    $y = 15x + 2$

Use a graph to solve the equation. Check your solution.

17. $-2x - 1 = -x - 3$

18. $3x + 1 = 4x + 2$

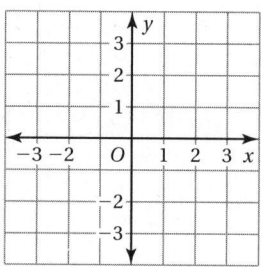

Graph the system of linear inequalities.

19. $y \geq -x + 1$
    $y < 2x + 1$

20. $x + 2y < -8$
    $x - 2y < 4$

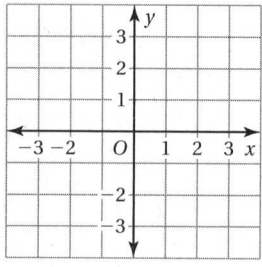

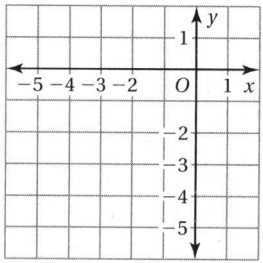

21. You have $3.75 to spend at a vending machine. You want to buy at least three health bars. Regular health bars cost $0.75 each and strawberry health bars cost $1.25 each.

   a. Write a system of linear inequalities that represents this situation.

   b. Is it possible to buy three regular health bars and one strawberry health bar? Justify your answer.

Answers

11. _____
12. _____
13. _____
14. _____
    _____
    _____
15. _____
    _____
    _____
16. _____
    _____
    _____
17. ___See left.___
    _____
18. ___See left.___
    _____
19. ___See left.___
20. ___See left.___
21. a. _____
    _____
    b. _____
    _____
    _____
    _____
    _____

46 Big Ideas Math Algebra 1
Assessment Book

Name_____ Date_____

Chapter 4 Standards Assessment

1. Which inequality is shown in the graph?

    A. $4x + 2y > 5$

    B. $4x + 2y < 5$

    C. $4x + 2y \geq 5$

    D. $4x + 2y \leq 5$

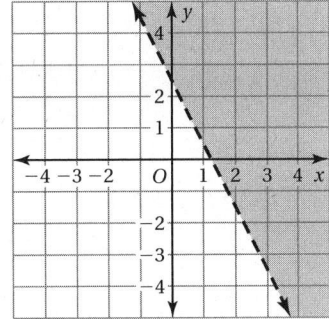

2. **GRIDDED RESPONSE** A middle school conducts a fire drill. The percent y (in decimal form) of students still inside x minutes after the fire alarm sounds is $y = -0.125x + 1$. After how many seconds are 75% of the students still inside?

3. The steps Andre took to solve the system of linear equations $y = 4x + 1$ and $y = 2x + 7$ are shown below. What should Andre change in order to correctly solve the system?

    $$4x + 1 = 2x + 7$$
    $$6x = 6$$
    $$x = 1$$

    F. The constants should combine to equal 8.

    G. The x-terms should combine to equal $2x$.

    H. The constants should combine to equal 2.

    I. The x-terms should combine to equal $-6x$.

4. The formula for average acceleration over a period of time is $A = \dfrac{v_f - v_0}{t}$. How can this formula be solved for final velocity v_f?

    A. Multiply both sides of the formula by t.

    B. Add v_0 to both sides of the formula.

    C. Subtract t from both sides of the formula then add v_0 to both sides of the formula.

    D. Multiply both sides of the formula by t then add v_0 to both sides of the formula.

Name _____ Date _____

Chapter 4 Standards Assessment (continued)

5. Which ordered pair is a solution to the system of linear inequalities below?

$$y < \frac{1}{4}x + 2$$
$$y \geq x - 1$$

   F. $(-4, 1)$ H. $(2, 1)$

   G. $(1, -1)$ I. $(6, 4)$

6. The town library is having a used book sale. The graph to the right can be used to find the total cost y to buy x books. The total cost includes the admission fee.

   What is the equation of the line shown?

   A. $y = x + 4$ C. $y = -x + 4$

   B. $y = x - 4$ D. $y = -x - 4$

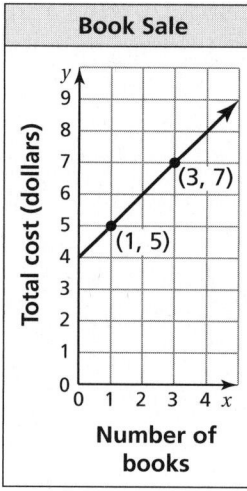

7. **SHORT RESPONSE** James and Max are saving their allowances to buy laptop computers. James has saved $30 already and earns a $5 allowance each week. Max has saved $10 already and earns a $10 allowance each week.

   Part A Write a system of equations that can represent this situation. Use x to represent the number of weeks and y to represent the total amount saved.

   Equation for James _____

   Equation for Max _____

   Part B After how many more weeks will James and Max have the same amount of money saved? Use your equations from Part A and the coordinate grid.

   Number of weeks _____

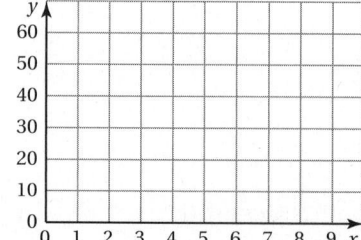

Chapter 4 Standards Assessment Item Analysis

1. **A.** Correct answer
   B. The student uses the wrong side of the line as the solution area.
   C. The student forgets that a dotted line signifies a non-inclusive inequality.
   D. The student uses the wrong side of the line as the solution area and forgets that a dotted line signifies a non-inclusive inequality.

2. Correct answer: 2
   Common error: The student adds 1 to the right side of the equation $-0.125x + 1 = 0.75$ instead of subtracting, leading to an incorrect answer of $x = -14$, which the student may change to 14.

3. **F.** The student adds 1 to each side, instead of subtracting 1.
   G. Correct answer
   H. The student divides each term by 4, but adds $\frac{1}{4}$ to each side instead of subtracting $\frac{1}{4}$.
   I. The student misapplies the sign rules of addition, thinking that 2 minus 4 equals -6.

4. **A.** The student takes one correct step then fails to completely solve for v_f.
   B. The student misunderstands what it means to solve for v_f and thinks $\frac{v_f - v_0}{t}$ can be separated using addition instead of multiplication.
   C. The student thinks $\frac{v_f - v_0}{t}$ can be separated using subtraction instead of multiplication.
   D. Correct answer

5. **F.** The student forgets that the "<" sign in the first inequality means points on that line are not solutions.
   G. The student only checks to see if the first inequality is satisfied.
   H. Correct answer
   I. The student confuses the meanings of the symbols in both inequalities.

Chapter 4 Standards Assessment Item Analysis
(continued)

6. **A.** Correct answer

   B. The student gets the correct slope but the wrong y-intercept. This student is not taking the information available from the graph and is making a sign error during arithmetic.

   C. The student gets the wrong slope but the correct y-intercept. This student is not taking the information available from the graph and is making a sign error during arithmetic.

   D. The student gets the wrong slope and the wrong y-intercept. This student is not taking the information available from the graph and is making a sign error during arithmetic.

7. **2 points** The student demonstrates a thorough understanding of setting up systems of equations and solving them by graphing. In Part A, the student writes the equations $y = 30 + 5x$ and $y = 10 + 10x$. In Part B, the student draws accurate graphs and obtains an answer of 4 weeks.

   1 point The student demonstrates limited understanding of setting up and solving systems of equations. For example, the system of equations has an error, the graph is drawn incorrectly, or the student cannot identify the correct point in Part B from a properly drawn graph.

   0 points The student provided no response, a completely incorrect or incomprehensible response, or a response that demonstrates insufficient understanding of setting up and solving systems of equations.

Chapter 4 Alternative Assessment

1. The equations $y = x + 4$ and $4y = -3x + 44$ are part of a system of linear equations that have one solution.

   a. Graph the equations to find the solution to this system of linear equations.

   b. Draw a line that represents another equation that is part of this system of linear equations and has the same solution. Let the y-intercept of this line be 0. Explain how you drew this line.

   c. Write the equation of the line you drew in part (b).

   d. Write two other equations that have the same solution and are in this system of linear equations. Graph these equations to show that they intersect at the same point as the others.

   e. Show algebraically that all five equations you have written are satisfied by the common solution.

   f. Choose two of the equations that are part of this system of linear equations and write a problem that can be solved with the two equations.

2. Consider the equations $y = 2x - 4$ and $5y = 3x + 15$.

   a. Explain why these equations form a system of linear equations.

   b. Solve the system of linear equations graphically and algebraically. Show your steps.

   c. Find the slope of each equation. Explain how the slope can be determined from the equation and from the graph of the equation.

   d. Identify the x- and y-intercept of each equation. Explain how they can be found algebraically and graphically.

   e. Rewrite each equation in standard form.

Name _____ Date _____

Chapter 4 Alternative Assessment Rubric

Score	Conceptual Understanding	Mathematical Skills	Work Habits
4	Shows complete understanding of: • solving systems of linear equations graphically and algebraically • identifying slope, x-intercept, and y-intercept of equations graphically and algebraically	Writes and graphs equation and interprets graph to answer all of Exercise 1. Correctly solves the system of linear equations graphically and algebraically. Identifies all features of both equations and rewrites both in standard form.	Answers all parts of both problems. All equations and graphs are written or drawn carefully and systematically. Work is very neat and well organized.
3	Shows nearly complete understanding of: • solving systems of linear equations graphically and algebraically • identifying slope, x-intercept, and y-intercept of equations graphically and algebraically	Writes and graphs equation and interprets graph to answer most of Exercise 1. Correctly solves the system of linear equations graphically or algebraically. Identifies some features of both equations and rewrites both in standard form.	Answers several parts of both problems. Most equations and graphs are written or drawn carefully and systematically. Work is neat and organized.
2	Shows some understanding of: • solving systems of linear equations graphically and algebraically • identifying slope, x-intercept, and y-intercept of equations graphically and algebraically	Writes and graphs equation and interprets graph to answer some of Exercise 1. Attempts to solve the system of linear equations graphically or algebraically. Identifies some features of both equations and rewrites one in standard form.	Answers some parts of both problems. Equations and graphs are written or drawn carelessly. Work is not very neat or organized.
1	Shows little understanding of: • solving systems of linear equations graphically and algebraically • identifying slope, x-intercept, and y-intercept of equations graphically and algebraically	Does not answer Exercise 1. Does not solve the system of linear equations either graphically or algebraically. Identifies no features of either equation and does not rewrite either in standard form.	Does not attempt any part of either problem. No equations or graphs are written or drawn. Work is sloppy and disorganized.

Name_____ Date_____

Chapter 5 Quiz
For use after Section 5.3

Find the domain and range of the function represented by the graph.

Answers

1. 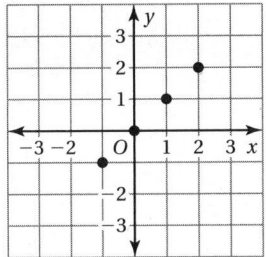   2.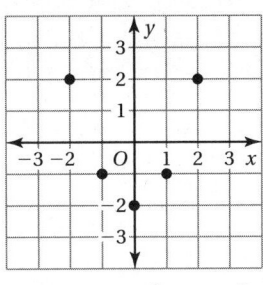

1. _____

_____

2. _____

_____

3. Graph the function. Is the domain discrete or continuous?

3. ___See left.___

Passengers, x	1	2	3	4
Cost, y	9	18	27	36

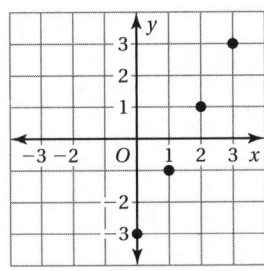

4. _____

5. _____

6. ___See left.___

Use the graph or table to write a linear function that relates y to x.

7. a._____

4. 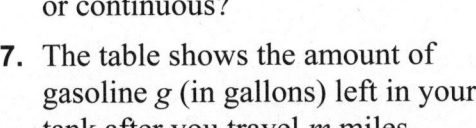   5.

x	−2	0	2	4
y	5	4	3	2

b._____

6. The function $y = 25 - 4x$ represents the amount y (in dollars) of money that you have after renting x movies. Graph the function using a domain of 0, 1, 2, 3, and 4. Is the domain discrete or continuous?

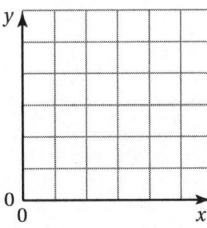

7. The table shows the amount of gasoline g (in gallons) left in your tank after you travel m miles.

   a. Write a linear function that relates the amount of gasoline to the traveling distance.

   b. How many gallons of gasoline are left after you drive 120 miles?

Miles, m	Gallons, g
0	20
20	19
40	18
60	17

Name _____ Date _____

Chapter 5 Quiz
For use after Section 5.6

Evaluate the function when x = –5, 0, and 10.

1. $f(x) = x - 4$
2. $g(x) = -3x + 5$
3. $h(x) = \dfrac{1}{5}x + 2$

Graph the function. Compare the graph to the graph of f(x) = 2x.

4. $g(x) = 2x + 3$

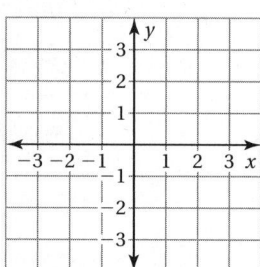

5. $h(x) = 2x - 2$

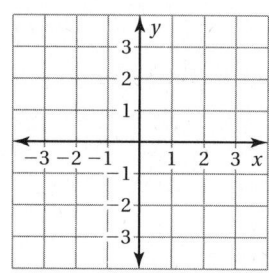

Does the table or graph represent a linear or nonlinear function? Explain.

6.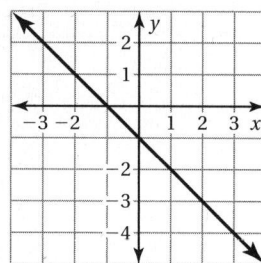

7.
x	y
2	5
4	17
6	37
8	65

Write an equation for the nth term of the arithmetic sequence. Then find a_{20}.

8. 11, 12, 13, 14, …

9. −3, −6, −9, −12, …

10. The function $f(x) = 6x - 40$ represents your profit (in dollars) after selling x bags of cookies at a bake sale. How much profit have you earned after selling 10 bags of cookies?

11. The table shows the value v (in thousands of dollars) of a house after t years. Does the table represent a linear or nonlinear function? Explain.

Year, t	0	1	2	3
Value, v	105	110	116	120

12. A restaurant charges $0.90 for a glass of milk and $0.50 for each refill. Write an equation for the nth term of the arithmetic sequence.

Answers

1. _____
   _____
   _____
2. _____
   _____
   _____
3. _____
   _____
   _____
4. ___See left.___
   _____
5. ___See left.___
   _____
6. _____
   _____
7. _____
   _____
8. _____
   _____
9. _____
   _____
10. _____
11. _____
    _____
12. _____

Name_____ Date_____

Chapter 5 Test A

1. Find the domain of the function represented by $(1, -5), (2, 3),$ and $(4, 7)$.

In Exercises 2–5, use the following information.

The number of shirts and shorts that you can buy with $30 is represented by the equation $10x + 5y = 30$. The table shows the number of shirts and shorts.

Shirts, x	0	1	2	3
Shorts, y	6	4	2	0

2. Write the equation in function form.

3. Find the range.

4. Graph the function.

5. Is the domain discrete or continuous?

Answers

1. _____
2. _____
3. _____
4. _____See left._____
5. _____
6. _____See left._____
7. _____
8. _____
9. _____
10. _____

6. Complete the input-output table for the function $y = 5x + 1$.

x	−1	0	1	2
y				

Is the domain discrete or continuous?

7.
Input Cats, x	Output Food (ounces), y
1	5.5
2	11
3	16.5

8.
Input Flour (tsp), x	Output Gravy (cups), y
2	1
4	2
6	3

Use the graph or table to write a linear function that relates y to x.

9.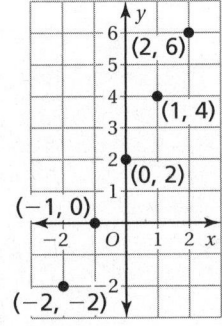

10.
x	−1	0	1	2
y	−4	0	4	8

Chapter 5 Test A (continued)

11. The function $D = 25 + 0.3x$ represents the daily rental charge D (in dollars) for x miles driven. Is the domain discrete or continuous?

Evaluate the function when $x = -4, 0,$ and 8.

12. $f(x) = x + 11$

13. $g(x) = -x + 6$

14. Compare the graph of $g(x) = 6x - 5$ to the graph of $f(x) = 6x$.

15. You earn $8 per hour working at a library. The function $p(x) = 8x$ represents the amount you earn for working x hours.

   a. You work 15 hours. How much do you earn?

   b. How many hours do you have to work to earn $200?

Does the graph represent a *linear* or *nonlinear* function? Explain.

16.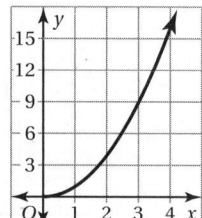

17.

18. The table shows the number y of muffins baked in x pans. What is the missing y-value that makes the table represent a linear function?

Pans, x	3	4	5
Muffins, y	18	?	30

Write an equation for the *n*th term of the arithmetic sequence. Then find a_{25}.

19. $8, 11, 14, 17, \ldots$

20. $-1, -6, -11, -16, \ldots$

Answers

11. _____
12. _____
    _____
13. _____
    _____
    _____
14. _____
    _____
15. a. _____
    b. _____
16. _____
    _____
    _____
17. _____
    _____
    _____
18. _____
19. _____
    _____
20. _____
    _____

Name_____ Date_____

Chapter 5 Test B

1. Find the domain and range of the function represented by the graph.

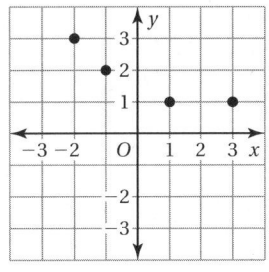

Answers

1. _____

_____

2. _____

3. _____

4. _____

5. __See left.__

6. _____

_____

7. _____

_____

8. _____

9. _____

10. a. __See left.__

    b. _____

In Exercises 2–4, use the table that shows the number of pens and notepads that you can buy with $12.

2. Find the domain.

3. Is the domain discrete or continuous?

Pens, x	0	4	8
Notepads, y	4	2	0

4. Find an equation in function form for this situation.

5. Complete the input-output table for the function $y = -1.2x + 4$.

x	−1	0	1	2
y				

6. You are putting wheels on skateboards. The function $y = 4x$ represents the number y of wheels that are needed for x skateboards. Is the number 15 in the domain? Explain.

7. The perimeter P of a square is a function of the length s of a side. Write a function for this perimeter. Is the function *linear* or *nonlinear*?

Use the graph or table to write a linear function that relates y to x.

8.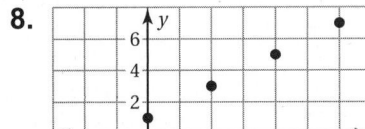

9.
x	−12	−6	0	6
y	6	3	0	−3

10. The function $y = 18 - 2x$ represents the number y of eggs left in a carton after cooking x omelets.

    a. Graph the function using a domain of 0, 1, 2, 3, and 4.

    b. Is the domain discrete or continuous?

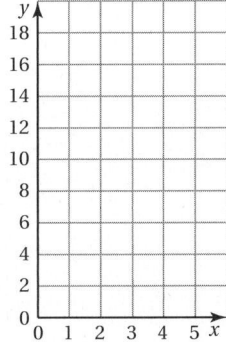

Chapter 5 Test B (continued)

Evaluate the function when $x = -5$, 0, and 10.

11. $f(x) = 2x - 6$

12. $g(x) = -\dfrac{1}{5}x - 7$

13. Compare the graph of $g(x) = 2x + \dfrac{1}{2}$ to the graph of $f(x) = 2x$.

14. You are packing candles in boxes. You can fit 15 candles in each box.

   a. Write a linear function using function notation that represents the number of candles that you pack into x boxes.

   b. How many boxes do you need to pack 75 candles?

15. The table shows the values y (in dollars) of Car A and Car B after x years of ownership. Which function represents a linear function: the function for *Car A*, for *Car B*, for *both*, or for *neither of them*?

Years, x	0	1	2	3
Value of Car A, y	24,000	20,000	16,000	12,000
Value of Car B, y	24,000	12,000	6000	3000

Does the equation represent a *linear* or *nonlinear* function? Explain.

16. $y = \dfrac{2}{x} + 1$

17. $y + 7 = 2x + 3y$

18. The table shows the cost y (in dollars) for x theater tickets. Find the missing y-value that makes the table represent a nonlinear function.

Tickets, x	2	4	6
Cost, y	26	?	78

Write an equation for the *n*th term of the arithmetic sequence. Then find a_{30}.

19. 1, 11, 21, 31, ...

20. −15, −10, −5, 0, ...

Answers

11. _____

_____

_____

12. _____

_____

_____

13. _____

_____

14. a. _____

b. _____

15. _____

16. _____

_____

17. _____

_____

18. _____

19. _____

20. _____

_____

Name _____ Date _____

Chapter 5 Standards Assessment

1. Diego is studying how hot dog sales at a ball park are related to attendance. To collect his data, Diego records the attendance at each game as an input value and hot dog sales at each game as an output value. Which statement best describes this situation?

   A. The domain is attendance and it is continuous.

   B. The domain is attendance and it is discrete.

   C. The domain is hot dog sales and it is continuous.

   D. The domain is hot dog sales and it is discrete.

2. **GRIDDED RESPONSE** Seven less than 9 times a number is equal to 20 more than the number. What is the number?

3. The results of a poll of registered voters in Ohio are shown in the table. The poll has a margin of error of ±4%. Which inequality describes the percent x of voters in Ohio that are affiliated with the Republican party?

Party Affiliation	
Democrat	44%
Republican	40%
Independent	16%

   F. $38 \le x \le 42$

   G. $40 \le x \le 44$

   H. $36 \le x \le 44$

   I. $40 \le x \le 48$

4. Which of the following are arithmetic sequences?

   I. 1, 4, 7, 10, …

   II. 12, 5, −2, −9, …

   III. 5, 10, 20, 40, …

   A. I

   B. I and II

   C. I, II, and III

   D. III

5. A painting's value increases at a rate of $5,000 per year. Ten years after it was purchased, the painting's value was $275,000. Which equation can be used to find v, the value of the picture in dollars, n years after it was purchased?

   F. $v = 280{,}000n$

   G. $v = 5{,}000n$

   H. $v = 5{,}000n + 275{,}000$

   I. $v = 5{,}000n + 225{,}000$

Name _____ Date _____

Chapter 5 Standards Assessment (continued)

6. Which method can you use to eliminate a variable from the following system of equations?

$$2x - 6y = 3$$
$$4x + y = -3$$

   A. Add the first equation to the second equation.

   B. Subtract the first equation from the second equation.

   C. Add twice the first equation to the second equation.

   D. Subtract twice the first equation from the second equation.

7. Which of the following is a solution to $4x < -28$?

   F. $x = -8$ **H.** $x = -6$

   G. $x = -7$ **I.** $x = 0$

8. The profit y from selling x muffins can be represented by a linear function. The profit from selling 5 muffins is $4. The profit from selling 7 muffins is $8. What is the slope of the line represented by the data?

   A. $\dfrac{1}{2}$ **C.** $\dfrac{4}{5}$

   B. 1   **D.** 2

9. **EXTENDED RESPONSE** To mail a letter, it costs $0.45 for the first ounce and $0.20 for each additional ounce. If a letter's weight is not a whole number of ounces, the weight is rounded up.

   Part A   Complete the input-output table.

Weight (ounces)	1	2	3	4	5
Cost (dollars)	$0.45				

   Part B   Identify the domain and range in the table.

   Domain _____ Range _____

   Part C   Is the domain continuous or discrete? Explain your reasoning.

   Part D   Let w be the weight of a letter and c be the cost of mailing the letter. Write an equation that models this situation.

   Equation _____

Chapter 5 Standards Assessment Item Analysis

1. **A.** The student mistakes attendance as continuous.
   B. Correct answer
   C. The student mistakes the range for the domain and mistakes discrete for continuous.
   D. The student mistakes the range for the domain.

2. Correct answer: 3.375
   Common error: The student adds x to $9x$ instead of subtracting or subtracts 7 from 20 instead of adding 7 to 20.

3. **F.** The student thinks a $\pm 4\%$ margin of error means the difference between the upper and lower bounds of the inequality must be 4%.
   G. The student only finds the upper limit of the inequality and uses the given percent for the lower limit.
   H. Correct answer
   I. The student uses the poll number for Democrat instead of Republican.

4. **A.** The student fails to recognize the common difference of terms in sequence II.
   B. Correct answer
   C. The student thinks the terms in sequence III have a common difference.
   D. The student misunderstands arithmetic sequences and thinks their consecutive terms have a common factor instead of a common difference.

5. **F.** The student adds 5,000 and 275,000.
   G. The student ignores the fixed term (initial value of the painting).
   H. The student uses the value after 10 years instead of the initial value.
   I. Correct answer

6. **A.** The student forgets that equations need variables with opposite coefficients to eliminate by addition or the student eliminates the constant instead of a variable.
   B. The student forgets that equations need variables with equal coefficients to eliminate by subtraction or incorrectly tries to eliminate the constant instead of a variable.
   C. The student forgets that equations need variables with opposite not equal coefficients to eliminate by addition.
   D. Correct answer

Chapter 5 Standards Assessment Item Analysis (continued)

7. **F.** Correct answer
   G. The student confuses "less than" with "less than or equal to."
   H. The student switches the inequality sign when dividing by 4 to solve for x or substitutes -6 into the inequality and confuses the signs.
   I. The student ignores negative numbers and picks zero because it is less than any positive number.

8. **A.** The student either found the change in x over the change in y or wrote the coordinates as (y, x).
   B. The student found the sum of the y-values over the sum of the x-values.
   C. The student only used the point $(5, 4)$ to find a slope of $\frac{y}{x}$.
   D. Correct answer

9. **4 points** The student demonstrates a thorough understanding of function concepts and how they apply to the application in this problem. Each part is answered correctly and clearly. In Part A, the costs are $0.65, $0.85, $1.05, and $1.25. In Part B, the domain is 1, 2, 3, 4, and 5, and the range is $0.45, $0.65, $0.85, $1.05, and $1.25. In Part C, the domain is discrete because the weight is rounded up to whole ounces. In Part D, the equation is $c = 0.2w + 0.25$.

   3 points The student demonstrates an essential but less than thorough understanding of function concepts and how they apply to the application in this problem. For example, the work done in Part C might be somewhat muddled.

   2 points The student demonstrates a partial understanding of function concepts and how they apply to the application in this problem. The student's work and explanations demonstrate a lack of essential understanding. For example, Part C may show muddled work, domain and range may be confused or misunderstood, and the equation in Part D could be incorrect.

   1 point The student demonstrates limited understanding. The student's response is incomplete and exhibits many flaws.

   0 points The student provides no response, a completely incorrect or incomprehensible response, or a response that demonstrates insufficient understanding of function concepts and how they apply to the application in this problem.

Name_____ Date _____

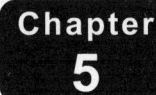 **Alternative Assessment**

1. **a.** Copy and complete each table.

$y = x + 2$	
x	y
2	
4	
6	
8	

$y = x - 2$	
x	y
2	
4	
6	
8	

$y = \dfrac{2}{x}$	
x	y
2	
4	
6	
8	

$y = \dfrac{x}{2}$	
x	y
2	
4	
6	
8	

$x = 2x$	
x	y
2	
4	
6	
8	

  b. Graph each function.

  c. Tell whether each function represents a *linear* or *nonlinear* function. Explain.

  d. Find the domain and range of each function.

2. Each table represents a line on the graph.

Table 1	
x	y
0	1
1	3
2	5
3	7

Table 2	
x	y
0	0
1	3
2	6
3	9

Table 3	
x	y
0	2
1	3
2	4
3	5

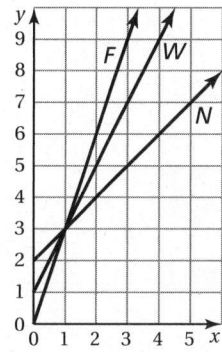

  a. Identify the table that represents each line on the graph.

  b. Write a linear function for each line that relates *y* to *x*.

  c. Choose one function and write a short word problem that can be represented by the table of that function with a discrete domain.

  d. Choose another function and write a short word problem that can be represented by the table of that function with a continuous domain.

Name_____ Date_____

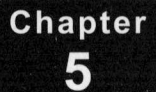 **Alternative Assessment Rubric**

Score	Conceptual Understanding	Mathematical Skills	Work Habits
4	Shows complete understanding of: • graphing functions • linear and nonlinear functions • domain and range • writing linear functions	Completes all tables correctly. Writes and graphs all functions correctly. Finds the domain and range of all functions in Exercise 1.	Answers all parts of all questions. All calculations are done carefully. Work is neat and well organized.
3	Shows nearly complete understanding of: • graphing functions • linear and nonlinear functions • domain and range • writing linear functions	Completes most tables correctly. Writes and graphs most functions correctly. Finds the domain and range of most functions in Exercise 1.	Answers almost all questions. Most calculations are done carefully. Most of the work is neat and well organized.
2	Shows some understanding of: • graphing functions • linear and nonlinear functions • domain and range • writing linear functions	Completes some tables correctly. Writes and graphs some functions correctly. Finds the domain and range of some functions in Exercise 1.	Answers some parts of all questions. Some calculations are done carefully. Some work is neat and well organized.
1	Shows little understanding of: • graphing functions • linear and nonlinear functions • domain and range • writing linear functions	Did not complete tables correctly. Did not write or graph functions correctly. Did not find the domain and range of functions in Exercise 1.	Answers few parts of the questions. No calculations are done carefully. All work is sloppy and disorganized.

Name_____ Date_____

Chapter 6 Quiz
For use after Section 6.3

Simplify the expression.

1. $\sqrt{96}$

2. $-\sqrt{\dfrac{17}{16}}$

3. $\dfrac{-10 - \sqrt{8}}{2}$

4. $\dfrac{7 + \sqrt{147}}{21}$

Evaluate the expression when $x = -6$, $y = 5$, and $z = 4$.

5. $\sqrt{x^2 z + xz^2}$

6. $\sqrt{3y^2 - xz}$

Simplify. Write your answer using only positive exponents.

7. $4^{-3} \cdot 4^8$

8. $(6d)^{-2}$

9. $\dfrac{n^2}{n^5}$

10. $\left(\dfrac{w^2}{3}\right)^3$

Simplify.

11. $\sqrt[4]{625}$

12. $64^{1/3}$

13. $9^{5/2}$

14. $16^{3/4}$

15. What is the area of the rectangular poster?

16. Find an example of two irrational numbers a and b for which $a \neq b$ and $\dfrac{a}{b}$ is rational.

17. An airplane travels a distance of $42x^5 y^4$ miles in $6xy^2$ hours. Find the average speed of the airplane.

18. The radius r of a sphere is given by the equation $r = \left(\dfrac{3V}{4\pi}\right)^{1/3}$, where V is the volume of the sphere. Find the radius of the sphere when $V = 36\pi$ cubic feet.

Answers

1. _____
2. _____
3. _____
4. _____
5. _____
6. _____
7. _____
8. _____
9. _____
10. _____
11. _____
12. _____
13. _____
14. _____
15. _____
16. _____
17. _____
18. _____

Name _____ Date _____

Chapter 6 Quiz
For use after Section 6.7

Does the table represent a *linear* or an *exponential* function? Explain.

1.
x	1	2	3	4
y	3	12	48	192

2.
x	0	3	6	9
y	1	7	13	19

Graph the function. Describe the domain and range.

3. $y = -3(2)^x$

4. $y = \left(\dfrac{2}{3}\right)^x$

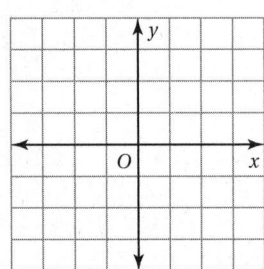
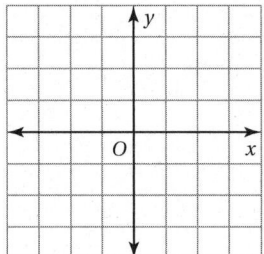

Solve the equation. Check your solution, if possible.

5. $9^{x+1} = 81^{2x+5}$

6. $125^{2x-3} = 5^{4x-5}$

Determine whether the table represents an *exponential growth function*, an *exponential decay function*, or *neither*.

7.
x	0	1	2	3
y	81	54	36	24

8.
x	0	2	4	6
y	9	36	144	576

Write the next three terms of the geometric sequence.

9. 2, 10, 50, 250, …

10. 896, −448, 224, −112, …

Write a recursive rule for the sequence.

11. 162, 54, 18, 6, …

12. −15, −7, 1, 9, …

13. The table shows the value of a car *t* years after it is purchased.

   a. Write a function that represents the value *y* of the car after *t* years.

   b. What is the value of the car after 7 years?

t	Value
0	$35,000
1	$30,100
2	$25,886
3	$22,261.96

Answers

1. _____

2. _____

3. ___See left.___

4. ___See left.___

5. _____

6. _____

7. _____

8. _____

9. _____

10. _____

11. _____

12. _____

13. a. _____

    b. _____

Name_____ Date_____

Chapter 6 Test A

Simplify the expression.

1. $\sqrt{60}$
2. $\sqrt{\dfrac{27}{16}}$
3. $\dfrac{6 - \sqrt{28}}{2}$
4. $\dfrac{15 + \sqrt{75}}{10}$

Simplify. Write your answer using only positive exponents.

5. $n^2 \cdot n^9$
6. $(5x)^3$
7. $\dfrac{d^{-4}}{d^3}$
8. $\left(\dfrac{x^5}{6}\right)^{-2}$

Simplify the expression.

9. $\sqrt[3]{125}$
10. $49^{1/2}$
11. $729^{2/3}$
12. $64^{4/3}$

Evaluate the function for the given value of *x*.

13. $y = 6^x; \ x = 2$
14. $f(x) = \dfrac{1}{9}(3)^x; \ x = 5$

15. Graph $y = 4^x + 2$. Describe the domain and range. Compare the graph to the graph of $y = 4^x$.

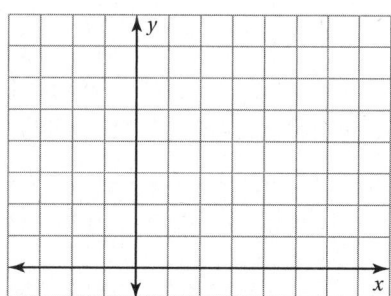

Write an exponential function represented by the table.

16.
x	0	1	2	3
y	1	7	49	343

17.
x	0	1	2	3
y	−4	−12	−36	−108

Solve the equation. Check your solution, if possible.

18. $3^{2x+5} = 3^{-3x}$
19. $2^{3x} = \dfrac{1}{64}$

Answers

1. _____
2. _____
3. _____
4. _____
5. _____
6. _____
7. _____
8. _____
9. _____
10. _____
11. _____
12. _____
13. _____
14. _____
15. ___See left.___
_____
_____
_____
16. _____
17. _____
18. _____
19. _____

Name _____ Date _____

Chapter 6 Test A (continued)

Write and graph a function that represents the situation.

20. Your $840 annual bonus increases by 5% each year.

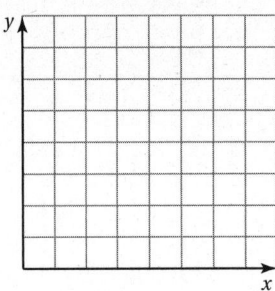

21. You deposit $3000 in an account that earns 2.5% annual interest compounded yearly.

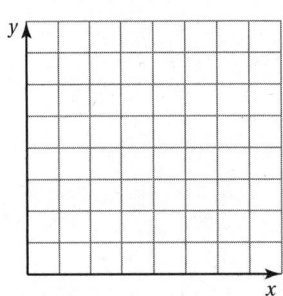

Answers

20. _____
    __See left.__

21. _____
    __See left.__

22. a. _____
    b. _____
    c. _____

23. _____
    _____

24. _____
    _____

25. _____

22. The table shows the value of a car over time.

Year, t	0	1	2	3
Value, y	$18,000	$15,300	$13,005	$11,054

a. Determine whether the table represents an *exponential growth function,* an *exponential decay function,* or *neither.*

b. The car loses 15% of its value every year. Write a function that represents the value y (in dollars) of the car after t years.

c. Use the function from part (b) to estimate the value of the car after 8 years. Round your answer to the nearest hundredth.

Write an equation for the nth term of the geometric sequence. Then find a_7.

23. 2, −6, 18, −54, …

24.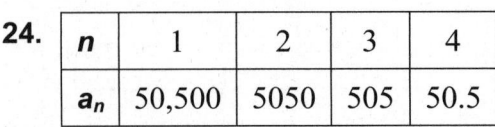

n	1	2	3	4
a_n	50,500	5050	505	50.5

25. The volume of the cube is 343 cubic yards. Find the dimensions of the cube.

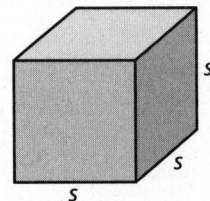

Name_____ Date_____

Chapter 6 Test B

Simplify the expression.

1. $-\sqrt{72}$

2. $\sqrt{\dfrac{175}{36}}$

3. $\dfrac{10 + \sqrt{44}}{4}$

4. $\dfrac{9 - \sqrt{27}}{-6}$

Simplify. Write your answer using only positive exponents.

5. $a^3 \cdot a^{-8}$

6. $(4m)^{-3}$

7. $\left(\dfrac{2f}{3f^2}\right)^{-2}$

8. $\dfrac{c^{-3} \cdot c}{c^7}$

Simplify the expression.

9. $\sqrt[3]{64}$

10. $121^{1/2}$

11. $16^{3/4}$

12. $243^{3/5}$

Evaluate the function for the given value of x.

13. $y = -3(4)^x$; $x = 2$

14. $f(x) = \dfrac{1}{4}(2)^x$; $x = 5$

15. Graph $y = 4^{x-2}$. Describe the domain and range. Compare the graph to the graph of $y = 4^x$.

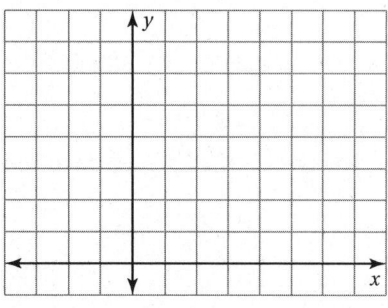

Write an exponential function represented by the table.

16.
x	0	1	2	3
y	2	12	72	432

17.
x	0	1	2	3
y	-5	-20	-80	-320

Solve the equation. Check your solution, if possible.

18. $5^{x+1} = 625$

19. $9^x = 27^{x+2}$

Answers

1. _____
2. _____
3. _____
4. _____
5. _____
6. _____
7. _____
8. _____
9. _____
10. _____
11. _____
12. _____
13. _____
14. _____
15. __See left.__
_____
_____
16. _____
17. _____
18. _____
19. _____

Name _____ Date _____

Chapter 6 Test B (continued)

Write and graph a function that represents the situation.

20. A population of 15,000 increases by 8.5% each year.

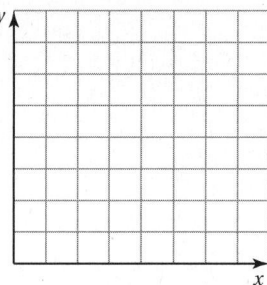

21. You deposit $350 in an account that earns 4.75% annual interest compounded yearly.

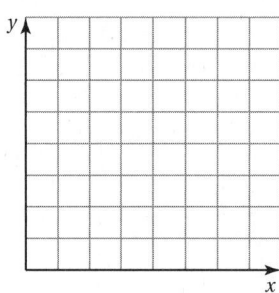

22. The function $P = 18,500(0.86)^t$ represents the value of a car after t years.

  a. Determine whether the table represents an *exponential growth function,* an *exponential decay function,* or *neither.*

  b. By what percent does the value of the car decrease each year?

  c. What is the value of the car after 8 years? Round your answer to the nearest hundredth.

Write an equation for the nth term of the geometric sequence. Then find a_7.

23. $-3, -6, -12, -24, \ldots$

24.
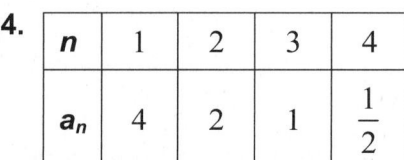

n	1	2	3	4
a_n	4	2	1	$\frac{1}{2}$

25. The volume of the cube is $\frac{125}{216}$ cubic yard. Find the dimensions of the cube.

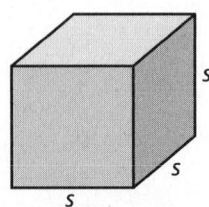

Answers

20. _____

    See left.

21. _____

    See left.

22. a. _____

    b. _____

    c. _____

23. _____

    _____

24. _____

    _____

25. _____

Name_____ Date_____

Chapter 6 Standards Assessment

1. Which statement about the following system of equations is true?

$$3x + 5y = 2$$
$$9x + 15y = 6$$

A. The system of equations has no solution.

B. The system of equations has infinitely many solutions.

C. The system of equations has exactly one solution, $(-1, 1)$.

D. The system of equations has exactly one solution, $\left(\dfrac{2}{3}, \dfrac{4}{5}\right)$.

2. GRIDDED RESPONSE What is the side length s of the cube in inches?

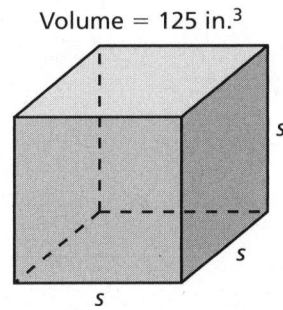

Volume = 125 in.³

3. The steps DeSean took to find the y-intercept of the equation are shown below. What should DeSean change in order to correctly find the y-intercept?

$$8x + 5y = 40$$
$$8x + 5(0) = 40$$
$$8x = 40$$
$$x = 5$$
y-intercept: $(0, 5)$

F. Let $x = 0$, instead of y.

G. Solve the equation for x.

H. The y-intercept should be written as $(5, 0)$.

I. Set both x and y equal to 0.

Chapter 6 Standards Assessment (continued)

4. The pressure on a scuba diver p (in pounds per square inch) can be modeled by the function $p = \frac{9}{20}x + 15$, where x is depth (in feet). At what depth, to the nearest foot, is the pressure equal to 130 pounds per square inch?

   A. 52 feet
   C. 256 feet
   B. 74 feet
   D. 322 feet

5. Linda invests $5000 in a money market account. The balance of the account in dollars after t years can be represented by the function $y = 5000(1.07)^t$. What is the account's annual rate of growth?

   F. 0.07%
   H. 107%
   G. 7%
   I. 5000%

6. Which of the following is a solution to the equation $|4x - 3| = 5$?

   A. $x = -8$
   C. $x = -\frac{1}{2}$
   B. $x = -2$
   D. $x = \frac{1}{2}$

7. **SHORT RESPONSE** The table shows the number of people y infected with a virus after x days.

x	0	1	2	3	4
y	1	3	9	27	81

   Part A Does the table represent a *linear function* or an *exponential function*? Explain.

   Part B Write and graph a function that represents this situation.

Chapter 6 Standards Assessment Item Analysis

1. **A.** The student recognizes that the equations have the same slope and incorrectly concludes they are parallel.
   B. Correct answer
   C. The student checks the point and sees that it satisfies both equations but fails to consider that there may be more solutions.
   D. The student attempts to solve the system of equations and makes an arithmetic error.

2. Correct answer: 5
   Common error: The student finds the square root of 125 instead of the cube root.

3. **F.** Correct answer
   G. The student has misapplied slope-intercept form. Solving the equation for x and identifying the constant term, will yield the x-intercept.
   H. The student has misunderstood that the x-value needs to be set to 0.
   I. The student has misunderstood the concept of y-intercept.

4. **A.** The student multiplies by $\frac{9}{20}$ instead of $\frac{20}{9}$ when solving for x.
   B. The student substitutes 130 in for x and finds p instead of vice-versa.
   C. Correct answer
   D. The student adds 15 to 130 instead of subtracting when solving for x.

5. **F.** The student confuses the decimal form with the percent form of the rate of growth.
   G. Correct answer
   H. The student confuses the growth factor with the rate of growth.
   I. The student confuses the initial amount with the rate of growth.

6. **A.** The student multiplies by 4 instead of dividing when solving $4x - 3 = -5$.
   B. The student correctly finds the solution $x = 2$ and incorrectly assumes $x = -2$ is a solution because of the absolute value sign.
   C. Correct answer
   D. The student correctly finds the solution $x = -\frac{1}{2}$ and incorrectly assumes $x = \frac{1}{2}$ is a solution because of the absolute value sign.

Chapter 6 Standards Assessment Item Analysis
(continued)

7. **2 points** The student demonstrates a thorough understanding of recognizing, writing, and graphing exponential functions. In part A, the student should say that the function is exponential and note that each successive term increases by a constant factor. In part B, the student should write the function $y = 3^x$ and graph it with reasonable accuracy by drawing a smooth curve through the points given in the table.

   1 point The student's work demonstrates limited understanding of recognizing, writing, and graphing exponential functions. Either the student cannot explain how they know the graph is exponential or the student graphs the function incorrectly.

   0 points The student provides no response, a completely incorrect or incomprehensible response, or a response that demonstrates insufficient understanding of recognizing, writing, and graphing exponential functions.

Name_____ Date_____

Chapter 6 Alternative Assessment

1. For each expression, simplify. Then explain how you know that the expression is in the most simplified form.

   a. $\sqrt{90}$

   b. $-\sqrt{\dfrac{72}{289}}$

   c. $\dfrac{4 - \sqrt{96}}{2}$

   d. $x^7 \cdot x^{-2}$

   e. $\dfrac{y^3}{y^7}$

   f. $\left(d^2\right)^{-3}$

   g. $\sqrt[3]{675}$

   h. $125^{4/3}$

2. The number of flowers in your garden each year is shown in the table.

Time (in years)	1	2	3	4
Number of flowers	24	36	54	81

   a. Does the table represent a *linear* or *exponential* function? Explain.

   b. Write the function represented by the table.

   c. Find the number of flowers in the garden after 8 years.

   d. Graph the function from part (b).

   e. Describe the domain and range.

   f. Find and interpret the *y*-intercept.

Name _____ Date _____

Chapter 6 — Alternative Assessment Rubric

Score	Conceptual Understanding	Mathematical Skills	Work Habits
4	Shows complete understanding of: • simplifying expressions with square roots and exponents • writing and graphing exponential growth functions	Simplified all expressions correctly and explained how to know when an expression is simplified. Correctly wrote, graphed, and answered questions about an exponential growth function.	Answers all parts of both questions. All calculations are done carefully. All work is neat and well organized.
3	Shows nearly complete understanding of: • simplifying expressions with square roots and exponents • writing and graphing exponential growth functions	Simplified most expressions correctly and explained how to know when an expression is simplified. Wrote, graphed, and answered questions about an exponential growth function, but made 1 or 2 errors.	Answers almost all parts of both questions. Most of the calculations are done carefully. Most of the work is neat and well organized.
2	Shows some understanding of: • simplifying expressions with square roots and exponents • writing and graphing exponential growth functions	Simplified some expressions correctly and explained how to know when an expression is simplified. Wrote, graphed, and answered questions about an exponential growth function, but made errors.	Answers some parts of both questions. Some calculations are done carefully. Some work is neat and well organized.
1	Shows little understanding of: • simplifying expressions with square roots and exponents • writing and graphing exponential growth functions	Simplified none of the expressions correctly or explained how to know when an expression is simplified. Did not write, graph, or answer any questions about an exponential growth function.	Answers few parts of both questions. No calculations are done carefully. All work is sloppy and disorganized.

Name_____ Date_____

Chapter 7 Quiz
For use after Section 7.4

Find the degree of the monomial.

1. 13
2. $-9r^2$
3. $2k^5 j$

Write the polynomial in standard form. Identify the degree and classify the polynomial by the number of terms.

4. $5y^4 + 4y^6$
5. $2r - 9 + r^3$

Find the sum or difference.

6. $(2k^4 + 3k) + (5k^4 - 6)$

7. $(2p^2 - 5p + 4) + (5p^2 + 2p + 1)$

8. $(3m^2 + 6m - 4) - (2m + 9)$

9. The cost (in dollars) of making x flip flops is represented by $40 + 4x$. The cost (in dollars) of making x sandals is represented by $70 + 8x$.

    a. Write a polynomial that represents the total cost of making x sandals and x flip flops.

    b. What is the total cost of making 15 sandals and 15 flip flops?

Find the product.

10. $(a - 1)(a - 5)$
11. $(3g - 4)(2g + 7)$
12. $(b + 9)(b - 2)$
13. $(5w + 2)(4w + 1)$
14. $(m - 7)(m + 7)$
15. $(3v + 8)^2$

16. A garden is extended on two sides.

    a. The area of the garden after the extension is represented by $(x + 12)^2$. Find this product.

    b. Use the polynomial in part (a) to find the area of the garden when $x = 8$. What is the area of the extension?

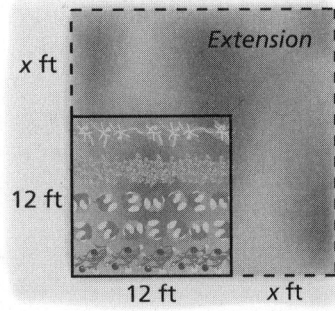

Answers

1. _____
2. _____
3. _____
4. _____
   _____
5. _____
   _____
   _____
6. _____
7. _____
8. _____
9. a._____
   b._____
10. _____
11. _____
12. _____
13. _____
14. _____
15. _____
16. a._____
    b._____

Chapter 7 Quiz
For use after Section 7.9

Factor the polynomial.

1. $3p - 18$
2. $33b^2 + 22$
3. $k^2 + 5k + 4$
4. $p^2 - 3p - 54$
5. $4m^2 + 14m + 6$
6. $7a^2 + 12a - 4$
7. $h^2 + 20h + 100$
8. $z^2 - 12z + 36$

Solve the equation.

9. $2t(t + 6) = 0$
10. $g^2 + 9g + 18 = 0$
11. $2a^2 + 6a = 20$
12. $5c^2 + 5c - 60 = 0$
13. $x^2 - x - 42 = 0$
14. $c^2 + 5c - 36 = 0$
15. $w^2 - 144 = 0$
16. $9v^3 = -45v^2$

Answers

1. _____
2. _____
3. _____
4. _____
5. _____
6. _____
7. _____
8. _____
9. _____
10. _____
11. _____
12. _____
13. _____
14. _____
15. _____
16. _____
17. _____
18. _____
19. _____

17. An arch of balloons is tied to a frame. The frame can be modeled by $y = -\dfrac{5}{8}(x - 2)(x - 10)$, where x and y are measured in feet. The x-axis represents the ground. Find the width of the frame at ground level.

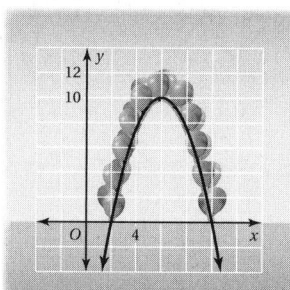

18. Find the dimensions of the rectangle when the area is 32 square inches.

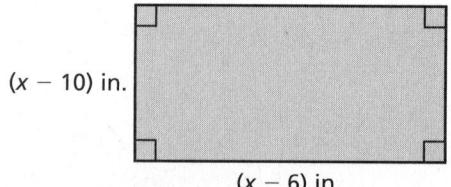

19. A ball is dropped from a height of 225 feet. The ball's height y (in feet) after t seconds can be modeled by $y = 225 - 16t^2$. After how many seconds does the ball hit the ground?

Name_____ Date_____

Chapter 7 Test A

Write the polynomial in standard form. Identify the degree and classify the polynomial by the number of terms.

1. $8z + 3z^2$
2. $6 - 3w - w^4$
3. $7m^5$

Find the sum or difference.

4. $(3q - 2) + (6q + 9)$
5. $(u^2 + 4u) - (8u - 2)$
6. $(3a^2 - 7ab) + (a^2 + 2ab - 6b^2)$

Find the product.

7. $(g + 6)(g + 7)$
8. $(b + 9)(b - 2)$
9. $(6 - 5t)(9 - 2t)$
10. $(3w + 4)(4w - 8)$
11. $(m - 7)(m + 7)$
12. $(3v + 8)^2$

Write a polynomial that represents the area of the figure.

13.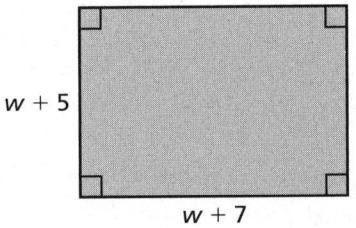
(rectangle with sides $w + 5$ and $w + 7$)

14.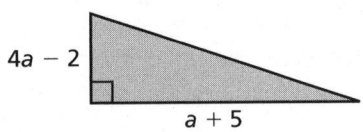
(right triangle with legs $4a - 2$ and $a + 5$)

15. You design the wooden poster frame and paint the front surface.

   a. Write a polynomial that represents the area of wood you paint.

   b. You design the frame to display a 16-inch by 20-inch poster. How much wood do you paint?

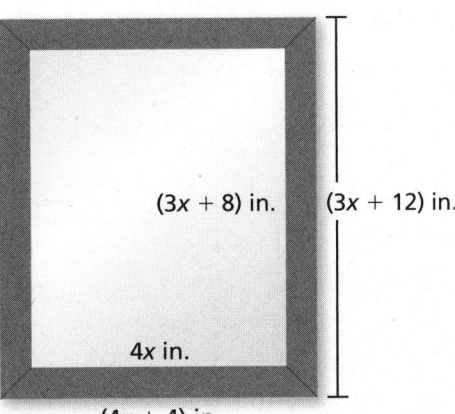
(Frame with outer dimensions $(3x + 12)$ in. by $(4x + 4)$ in. and inner opening $(3x + 8)$ in. by $4x$ in.)

Answers

1. _____
   _____
   _____
2. _____
   _____
   _____
3. _____
   _____
4. _____
5. _____
6. _____
7. _____
8. _____
9. _____
10. _____
11. _____
12. _____
13. _____
14. _____
15. a. _____
    b. _____

Chapter 7 Test A (continued)

Factor the polynomial.

16. $5c^4 + 60c$

17. $y^2 - 10y + 21$

18. $a^2 + 2a - 15$

19. $m^2 - 3m - 28$

20. $w^2 - 81$

21. $6x^2 - 7x - 3$

Solve the equation.

22. $(w + 3)(w - 10) = 0$

23. $3n^2 + 12n = 0$

24. $6q^2 = 24q$

25. $v^2 - 13v + 22 = 0$

26. $t^2 - 5 = 16 - 4t$

27. $4x^2 + 14x - 8 = 0$

28. $25u^2 - 36 = 0$

29. $z^2 - 24z + 144 = 0$

30. A basketball bounces off the ground. The ball's height y (in feet) after t seconds can be modeled by $y = -16t^2 + 20t$.

    a. How many seconds is the basketball in the air?

    b. The basketball reaches its maximum height after 0.625 second. What is the maximum height of the bounce?

31. A company's profit (in millions of dollars) can be represented by $x^2 - 5x + 6$, where x is the number of years since the company started. When did the company have a profit of $2 million?

32. You are adding an addition to your patio. The area (in square feet) of the addition can be represented by $k^2 - 3k - 10$.

    a. The area of the patio before the addition was 50 square feet. Find k.

    b. Find the area of the addition and the area of the entire patio after the addition.

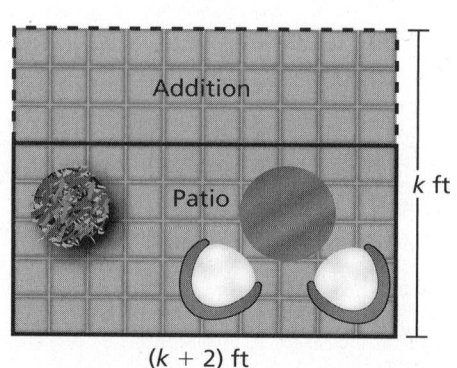

$(k + 2)$ ft

Answers

16. _____
17. _____
18. _____
19. _____
20. _____
21. _____
22. _____
23. _____
24. _____
25. _____
26. _____
27. _____
28. _____
29. _____
30. a. _____
    b. _____
31. _____
32. a. _____
    b. _____

Name_____ Date_____

Chapter 7 Test B

Write the polynomial in standard form. Identify the degree and classify the polynomial by the number of terms.

1. $12 - 7h^3$
2. $3y^2 - 6y^8 - 4$
3. $1.6d^2$

Find the sum or difference.

4. $(t^2 - 9t + 3) + (4t^2 + t + 5)$
5. $(6w - 3w^3) - (8w^3 - 4w)$
6. $(3r^2 + 7s^2) - (r^2 + 2rs + 13s^2)$

Find the product.

7. $(n + 9)(n - 3)$
8. $(p - 6)(p + 8)$
9. $(b - 7)(5b - 3)$
10. $(4s + 8)(4s - 8)$
11. $(k - 6)^2$
12. $(5y + 11)^2$

13. You want to cut out a parallelogram from a rectangle for a unique frame in your scrapbook.

    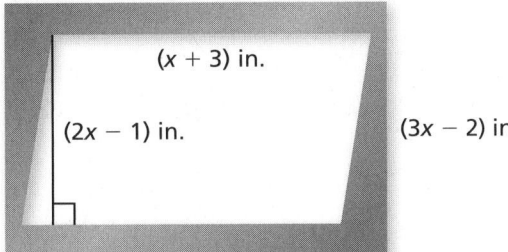

    $(x + 3)$ in.
    $(2x - 1)$ in.
    $(3x - 2)$ in.
    $(4x - 2)$ in.

    a. Write a polynomial that represents the area of the frame.

    b. What is the area of the frame when $x = 2$?

14. You are saving for a bicycle that costs $120. You deposit $75 in an account that earns interest compounded annually. The expression $75(1 + r)^2$ represents the balance after 2 years, where r is the annual interest rate in decimal form.

    a. Write a polynomial that represents the balance of your account.

    b. What is the balance of your account when the interest rate is 10%?

    c. How much more money do you need to buy the bicycle?

Answers

1. _____
   _____
   _____
2. _____
   _____
   _____
3. _____
   _____
4. _____
5. _____
6. _____
7. _____
8. _____
9. _____
10. _____
11. _____
12. _____
13. a. _____
    b. _____
14. a. _____
    b. _____
    c. _____

Name _____ Date _____

Chapter 7 Test B (continued)

Factor the polynomial.

15. $4w^3 - 8w$

16. $m^2 - 16m + 39$

17. $a^2 + 9a + 14$

18. $h^2 + 14h - 32$

19. $64 - 9a^2$

20. $3f^3 - 3f^2 - 36f$

Solve the equation.

21. $8x^2 + 24x = 0$

22. $8k^2 = 128k$

23. $c^2 - 12c + 27 = 0$

24. $f^2 - 22f = -72$

25. $d^2 + 11d - 26 = 0$

26. $4w^2 - 18w = 10$

27. $25v^2 = 4$

28. $s^2 + 121 = 22s$

29. A ball is thrown upward from a height of 80 feet. The ball's height y (in feet) after t seconds can be modeled by $y = -16t^2 + 64t + 80$.

   a. How long until the ball hits the ground?

   b. How long until the ball reaches the height from which it was thrown?

   c. The ball reaches its maximum height after 2 seconds. What is the maximum height of the ball?

30. A company's profit (in millions of dollars) can be represented by $x^2 - 6x + 12$, where x is the number of years since the company started. When did the company have a profit of $4 million?

31. You are building an addition to your deck. The area (in square yards) of the original deck can be represented by $y^2 + 2y - 15$.

   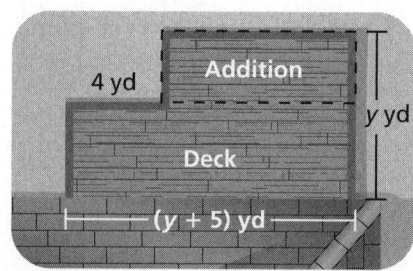

   a. The area of the addition is 24 square yards. Find y.

   b. Find the area of the entire deck after the addition.

Answers

15. _____
16. _____
17. _____
18. _____
19. _____
20. _____
21. _____
22. _____
23. _____
24. _____
25. _____
26. _____
27. _____
28. _____
29. a. _____
    b. _____
    c. _____
30. _____
31. a. _____
    b. _____

Name_____ Date_____

Chapter 7 Standards Assessment

1. Which polynomial represents the area of the parallelogram?

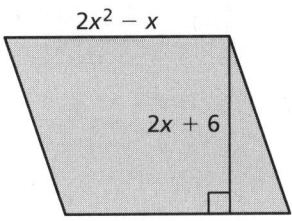

   A. $2x^3 + 14x^2 - 6x$

   C. $4x^3 + 14x^2 - 6x$

   B. $2x^3 + 10x^2 - 6x$

   D. $4x^3 + 10x^2 - 6x$

2. Which graph represents the inequality below?

   $x - 2 > 0$

   F.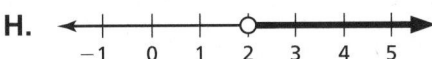

   H. (number line with open circle at 2, shaded right)

   G. (number line with closed circle at 2, shaded left)

   I.

3. In 24 years, Elliot's age will be 3 times what it is today. Let a represent Elliot's age today. Which equation can be used to find a?

   A. $24 = 3a$

   C. $24 - a = 3a$

   B. $a - 24 = 3a$

   D. $a + 24 = 3a$

4. **GRIDDED RESPONSE** What value of x makes the equation below true?

   $64^{x+2} = 4^{5x}$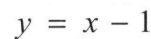

5. Which system of linear equations is shown in the graph?

   F. $y = \dfrac{5}{2}x + 1$
   $y = x - 5$

   H. $y = \dfrac{2}{5}x - 5$
   $y = x + 1$

   G. $y = \dfrac{5}{2}x - 5$
   $y = x + 1$

   I. $y = \dfrac{5}{2}x + 2$
   $y = x - 1$

   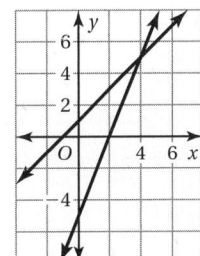

Chapter 7 Standards Assessment (continued)

6. The steps Audrey took to find the difference between $3x^3 + 7x^2 - x$ and $x^3 - 4x + 2$ are shown below. What should Audrey change in order to have the correct difference?

$$\begin{array}{r} (3x^3 + 7x^2 - x) \\ - (x^3 - 4x + 2) \\ \hline 2x^3 + 11x^2 - x - 2 \end{array}$$

   A. Nothing; she found the difference correctly.

   B. The x^2-coefficient should be 7 and the x-coefficient should be 3.

   C. The x^2-coefficient should be 7 and the x-coefficient should be -5.

   D. The x^2-coefficient should be 3.

7. Use the table below. Which linear function relates y to x?

x	1	3	5	7	9
y	11	9	7	5	3

   F. $y = 11x$ H. $y = -x + 12$

   G. $y = -2x + 13$ I. $y = x - 2$

8. **EXTENDED RESPONSE** The projected profit (in thousands of dollars) of a new technology company in year t can be modeled by the polynomial $2t^2 - 9t - 18$.

   Part A   Factor the polynomial and find when the company will break even (have zero profit). Show all work necessary to justify your answer.

   Part B   The projected profit (in thousands of dollars) in year t of a newly developed line of business can be modeled by the monomial t^3. Write a polynomial that models the company's total profit including this new line of business.

   Total Profit _____

   Part C   Factor the polynomial from Part B and find how many years sooner the company will break even with the new line of business. Show all work necessary to justify your answer.

Chapter 7 Standards Assessment Item Analysis

1. **A.** The student incorrectly multiplies $2x^2$ and $2x$ and incorrectly adds $12x^2$ and $-2x^2$.
   B. The student incorrectly multiplies $2x^2$ and $2x$.
   C. The student incorrectly adds $12x^2$ and $-2x^2$.
   D. Correct answer

2. **F.** The student confuses the meaning of >.
   G. The student confuses the meaning of > and incorrectly includes the number 2.
   H. Correct answer
   I. The student incorrectly includes the number 2.

3. **A.** The student represents Elliot reaching the age of 24 instead of aging 24 years.
   B. The student misrepresents Elliot aging 24 years.
   C. The student misrepresents Elliot aging 24 years.
   D. Correct answer

4. Correct answer: 3
   Common error: The student equates $x + 2$ and $5x$ and gets $\frac{1}{2}$ for the answer.

5. **F.** The student matches one equation's slope with the other's y-intercept and vice-versa.
   G. Correct answer
   H. The student uses the reciprocal of the slope for the first equation.
   I. The student uses the x-intercepts instead of the y-intercepts to write the equations in slope-intercept form.

6. **A.** The student subtracts $-4x$ from $7x^2$ instead of subtracting $-4x$ from $-x$.
   B. Correct answer
   C. The student subtracts $4x$ from $-x$ instead of subtracting $-4x$ from $-x$.
   D. The student subtracts $4x$ from $7x^2$ instead of subtracting $-4x$ from $-x$.

7. **F.** The student picks the simplest linear function that fits the first point in the table.
   G. The student incorrectly calculates the slope, writes an equation, and tests only the first point.
   H. Correct answer
   I. The student analyzes the relationship among the y-values and selects an equation that most closely expresses that relationship.

Chapter 7 Standards Assessment Item Analysis
(continued)

8. **4 points** The student demonstrates a thorough understanding of factoring $ax^2 + bx + c$, adding monomials to polynomials, factoring cubic polynomials by grouping, and solving polynomial equations in factored form. Each part is answered correctly and clearly. For Part A, the student factors the polynomial as $(2t + 3)(t - 6)$, gets the answer $t = 6$, and excludes the negative root. For Part B, the student writes the polynomial $t^3 + 2t^2 - 9t - 18$. For Part C, the student factors the polynomial as $(t^2 - 9)(t + 2)$, gets the answer $t = 3$, excludes negative roots, and gets a final answer of 3 years.

3 points The student demonstrates an essential but less than thorough understanding of factoring $ax^2 + bx + c$, adding monomials to polynomials, factoring cubic polynomials by grouping, and solving polynomial equations in factored form. For example, negative roots are included as answers.

2 points The student demonstrates a partial understanding of factoring $ax^2 + bx + c$, adding monomials to polynomials, factoring cubic polynomials by grouping, and solving polynomial equations in factored form. The student's work and explanations demonstrate a lack of essential understanding. For example, correct factoring techniques are attempted but the factors are incorrect.

1 point The student demonstrates limited understanding of factoring $ax^2 + bx + c$, adding monomials to polynomials, factoring cubic polynomials by grouping, and solving polynomial equations in factored form. The student's response is incomplete and exhibits many flaws.

0 points The student provides no response, a completely incorrect or incomprehensible response, or a response that demonstrates insufficient understanding of factoring $ax^2 + bx + c$, adding monomials to polynomials, factoring cubic polynomials by grouping, and solving polynomial equations in factored form.

Name_____ Date_____

 **Alternative Assessment**

1. Consider the polynomial $-9x^2 + 2x^3 - 5x$.

   a. Write the polynomial in standard form. Identify the degree and classify it by the number of terms.

   b. Add $3x^2 + 5$ to the polynomial.

   c. Subtract $x^2 + 4x - 1$ from the polynomial.

   d. Multiply the polynomial by $x + 1$.

   e. Divide the polynomial by $x - 3$.

   f. Factor the polynomial.

   g. Set the polynomial equal to zero and find the solutions to the equation.

2. Explain why $(a + b)^2 \neq a^2 + b^2$. Then choose values for a and b and substitute them into the expression.

3. The area of a square is represented by the expression $4x^2 - 12x + 9$.

   a. Find the expression that represents one side of the square.

   b. Find the expression that represents the perimeter of the square.

   c. The area of the square is 289 cubic feet. Find the perimeter.

Name_____ Date _____

Alternative Assessment Rubric

Score	Conceptual Understanding	Mathematical Skills	Work Habits
4	Shows complete understanding of: • adding, subtracting, multiplying, dividing and factoring polynomials • solving polynomial equations	Performed operations with polynomials, found the solutions of the polynomial in Exercise 1, explained why the square of a sum is not equal to the sum of the squares, factored the area of the square, and found the length of each side and the perimeter with no errors.	Answers all parts of each question. All calculations are done carefully. All work is neat and well organized.
3	Shows nearly complete understanding of: • adding, subtracting, multiplying, dividing and factoring polynomials • solving polynomial equations	Performed operations with polynomials, found the solutions of the polynomial in Exercise 1, explained why the square of a sum is not equal to the sum of the squares, factored the area of the square, and found the length of each side and the perimeter with 1 or 2 errors.	Answers almost all parts of each question. Most of the calculations are done carefully. Most of the work is neat and well organized.
2	Shows some understanding of: • adding, subtracting, multiplying, dividing and factoring polynomials • solving polynomial equations	Performed operations with polynomials, found the solutions of the polynomial in Exercise 1, explained why the square of a sum is not equal to the sum of the squares, factored the area of the square, and found the length of each side and the perimeter with many errors.	Answers some parts of each question. Some calculations are done carefully. Some work is neat and well organized.
1	Shows little understanding of: • adding, subtracting, multiplying, dividing and factoring polynomials • solving polynomial equations	Incorrectly performed operations with polynomials, found the solutions of the polynomial in Exercise 1, explained why the square of a sum is not equal to the sum of the squares, and found the dimensions of the square.	Answers few parts of each question. No calculations are done carefully. All work is sloppy and disorganized.

Big Ideas Math Algebra 1
Assessment Book

Name_____ Date_____

Chapter 8 Quiz For use after Section 8.3

Identify the characteristics of the graph of the quadratic function.

Answers

1. 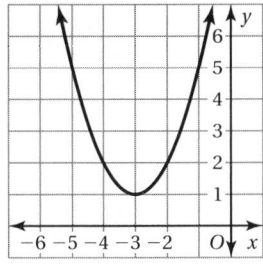   2.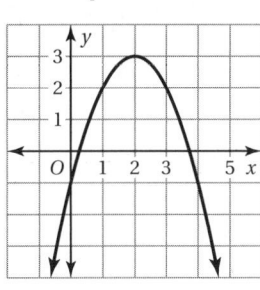

1. _____

_____

_____

_____

2. _____

Graph the function. Compare the graph to the graph of $y = x^2$.

3. $y = 3x^2$ 4. $y = -\dfrac{1}{2}x^2 + 4$

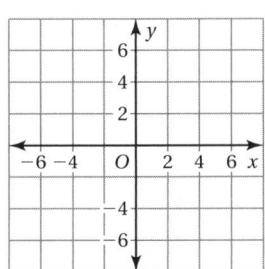   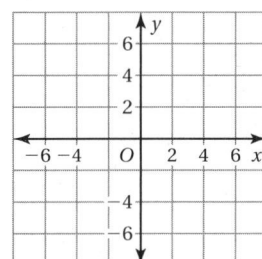

3. ____See left.____

4. ____See left.____

Graph the function. Identify the focus.

5. $y = -4x^2$ 6. $y = \dfrac{3}{8}x^2$

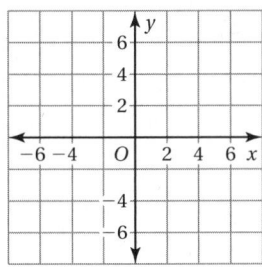   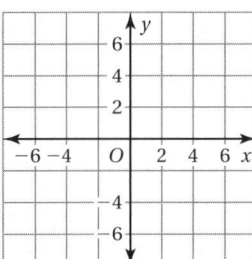

5. ____See left.____

6. ____See left.____

7. _____

8. _____

7. Write an equation of the parabola with focus (0, 4) and its vertex at the origin.

8. You build a satellite dish in the shape of a parabola, such that the feedhorn, located at the focus, is 16 inches from the vertex. Write an equation of the parabola with focus (0, 16) and its vertex at the origin.

Name _____ Date _____

Chapter 8 Quiz
For use after Section 8.5

Find (a) the axis of symmetry and (b) the vertex of the graph of the function.

1. $y = 3x^2 + 12x + 5$

2. $y = -\frac{1}{2}x^2 + 4x - 7$

Graph the function. Describe the domain and range.

3. $y = x^2 + 8x + 17$

4. $y = -3x^2 - 12x - 10$

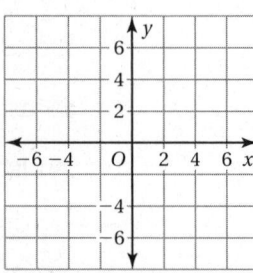
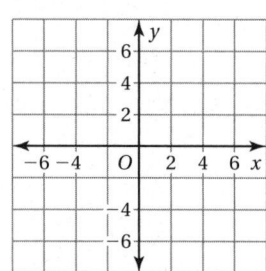

Tell whether the function has a minimum value or a maximum value. Then find the value.

5. $y = x^2 + 10x - 32$

6. $y = -\frac{1}{2}x^2 + 6x - 14$

Graph the function. Compare the graph to the graph of $y = x^2$.

7. $y = -(x - 4)^2 + 2$

8. $y = \frac{1}{3}(x + 1)^2 - 3$

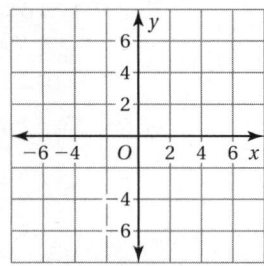
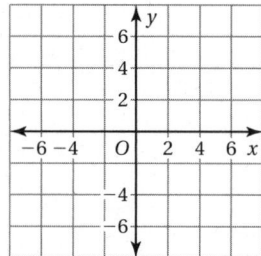

9. Tell whether the table of values represents a *linear*, an *exponential*, or a *quadratic* function. Then write an equation for the function using the form $y = mx + b$, $y = ab^x$, or $y = ax^2$.

x	−3	−2	−1	0	1
y	−4.5	−2	−0.5	0	−0.5

10. The function $f(t) = -16t^2 + 32t + 42$ gives the height h (in feet) of water from a fountain t seconds after it leaves the spout. When does the water reach its maximum height? What is the maximum height?

Answers

1. a. _____
   b. _____

2. a. _____
   b. _____

3. ___See left.___

4. ___See left.___

5. _____

6. _____

7. ___See left.___

8. ___See left.___

9. _____

10. _____

Name_____ Date_____

Chapter 8 Test A

Graph the function. Compare the graph to the graph of $y = x^2$.

Answers

1. $y = -4x^2$

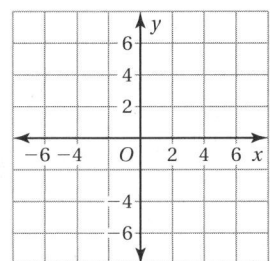

2. $y = \dfrac{2}{3}x^2$

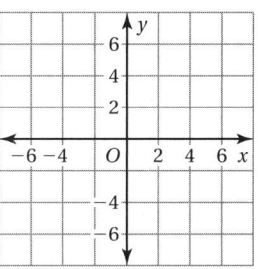

1. _____See left._____

2. _____See left._____

3. $y = x^2 - 5$

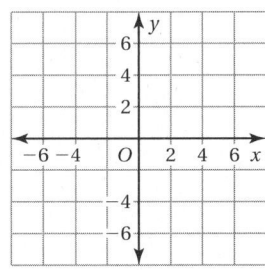

4. $y = -2x^2 + 3$

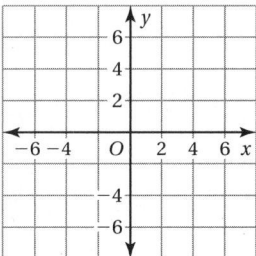

3. _____See left._____

4. _____See left._____

Graph the function. Identify the focus.

5. $y = \dfrac{1}{4}x^2$

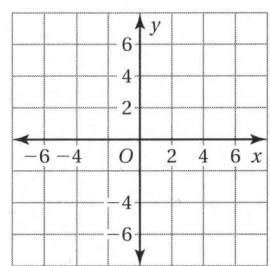

6. $y = -3x^2$

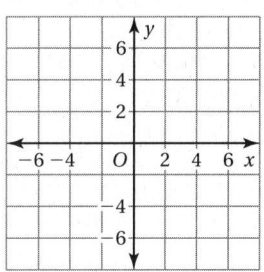

5. _____See left._____

6. _____See left._____

7. _____See left._____

Graph the function. Describe the domain and range.

7. $y = x^2 - 2x - 1$

8. $y = -2x^2 + 8x - 11$

8. _____See left._____

Copyright © Big Ideas Learning, LLC
All rights reserved.

Chapter 8 Test A (continued)

Graph the function. Compare the graph to the graph of $y = x^2$.

9. $y = (x + 3)^2$

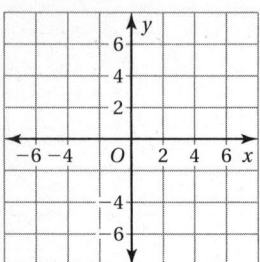

10. $y = -(x - 2)^2 + 1$

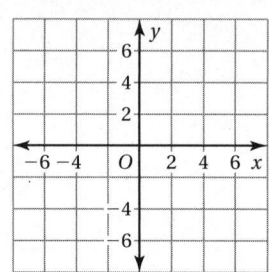

Answers

9. _____See left._____

_____

_____

10. _____See left._____

_____

_____

11. Write an equation of the parabola with focus $(0, 2)$ and its vertex at the origin.

12. An apple is dropped from a height of 64 feet. The function $h = -16t^2 + 64$ gives the height of the apple after t seconds.

   a. Graph the function.

   b. When does the apple hit the ground?

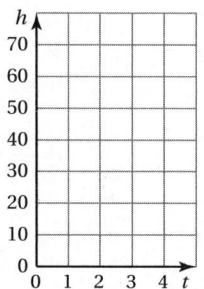

11. _____

12. a. _____See left._____

    b._____

13. _____

    _____

14. a._____

    b._____

    c._____

13. Tell whether the table of values represents a *linear*, an *exponential*, or a *quadratic* function. Then write an equation for the function using the form $y = mx + b$, $y = ab^x$, or $y = ax^2$.

x	1	2	3	4	5
y	0.5	2	4.5	8	12.5

14. The table shows the distance d (in miles) you bike each week.

Week, w	1	2	3	4
Distance, d	0.75	3	6.75	12

   a. Does a *linear*, an *exponential*, or a *quadratic* function represent the situation?

   b. Write a function that models the data.

   c. Use the function from part (b) to predict the amount you will bike in the seventh week.

Name_____ Date_____

Chapter 8 Test B

Graph the function. Compare the graph to the graph of $y = x^2$.

Answers

1. $y = 3x^2$
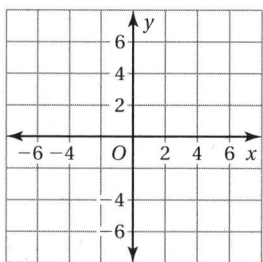

2. $y = -\dfrac{1}{2}x^2$
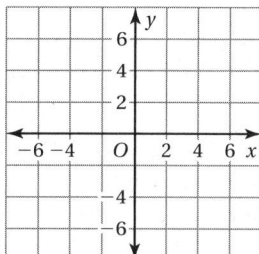

1. ___See left.___

2. ___See left.___

3. $y = \dfrac{1}{2}x^2 + 1$
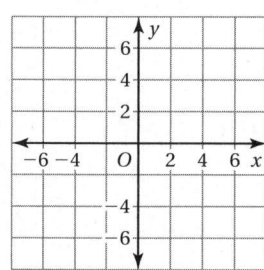

4. $y = -3x^2 + 5$
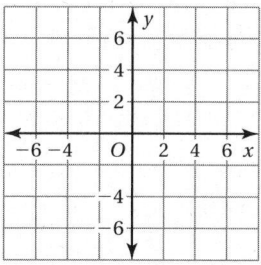

3. ___See left.___

4. ___See left.___

Graph the function. Identify the focus.

5. $y = \dfrac{5}{2}x^2$
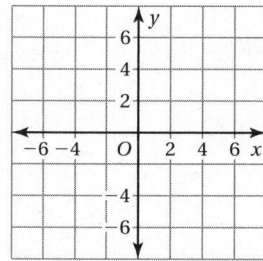

6. $y = -x^2$
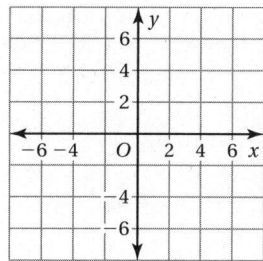

5. ___See left.___

6. ___See left.___

7. ___See left.___

Graph the function. Describe the domain and range.

7. $y = x^2 - 6x + 8$

8. $y = -\dfrac{1}{2}x^2 + 2x + 2$

8. ___See left.___

Name _____ Date _____

Chapter 8 Test B (continued)

Graph the function. Compare the graph to the graph of $y = x^2$.

9. $y = (x + 2)^2 - 4$

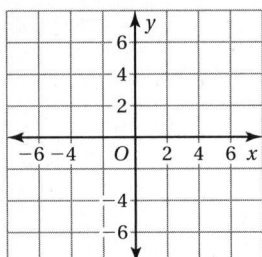

10. $y = -\dfrac{1}{2}(x - 1)^2 + 3$

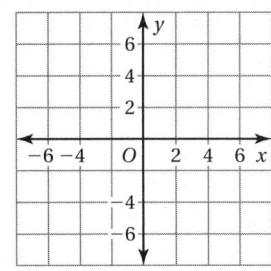

Answers

9. ___See left.___

10. ___See left.___

11. Write an equation of the parabola with focus $(0, -1.5)$ and its vertex at the origin.

12. An acorn is dropped from a height of 144 feet. The function $h = -16t^2 + 144$ gives the height of the acorn after t seconds.

    a. Graph the function.

    b. When does the acorn hit the ground?

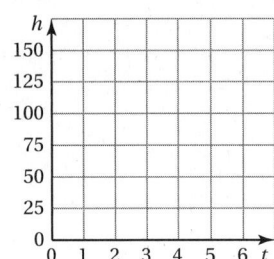

11. _____

12. a. ___See left.___

    b. _____

13. _____

14. a. _____

    b. _____

    c. _____

13. Tell whether the table of values represents a *linear*, an *exponential*, or a *quadratic* function. Then write an equation for the function using the form $y = mx + b$, $y = ab^x$, or $y = ax^2$.

x	−1	0	1	2	3
y	−2	0	−2	−8	−18

14. The table shows the amount a (in inches) of snowfall for each month.

Month, m	1	2	3	4
Amount, a	2	4	8	16

   a. Does a *linear*, an *exponential*, or a *quadratic* function represent the situation?

   b. Write a function that models the data.

   c. Use the function from part (b) to predict the amount of snowfall during the sixth month.

Chapter 8 Standards Assessment

1. Which expression is not equivalent to d^3?

   A. $\dfrac{d}{d^{-2}}$

   B. $\left(\dfrac{d^4}{d^3}\right)^{-3}$

   C. $\dfrac{(d^2)^4}{d^5}$

   D. $d^{-4}d^7$

2. Which table of values represents a quadratic function?

   F.
x	−2	−1	0	1	2
y	3	1	1	3	7

   G.
x	−2	−1	0	1	2
y	−16	−10	−4	2	8

   H.
x	0	1	2	3	4
y	0.5	1	2	4	8

   I.
x	0	1	2	3	4
y	−2	−3	−2	3	14

3. **GRIDDED RESPONSE** What is the 12th term divided by the 10th term for the sequence below?

   $$a_n = 5(3)^{n-1}$$

4. Which of the following is true for the graph?

   A. $h = -3; k = -1$

   B. $h = -3; k = 1$

   C. $h = 3; k = -1$

   D. $h = 3; k = 1$

   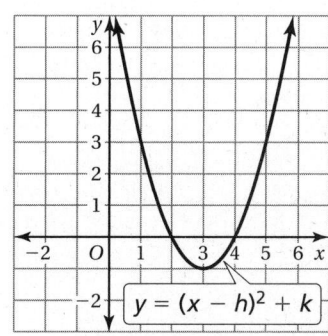

Chapter 8 Standards Assessment (continued)

5. Which linear equation can be solved with the graph?

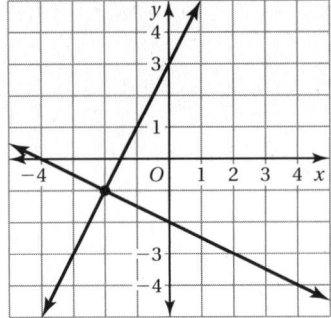

F. $x + 3 = -\dfrac{1}{3}x - 2$

G. $2x + 2 = -\dfrac{1}{2}x - 1$

H. $2x + 3 = -\dfrac{1}{2}x - 2$

I. $2x + 3 = -\dfrac{1}{3}x - 1$

6. Which expression is equivalent to $4h^2 - 25$?

A. $(2h + 5)(2h - 5)$ C. $(2h + 5)^2$

B. $(4h + 25)(4h - 25)$ D. $(2h - 5)^2$

7. Which of the following expressions is a polynomial?

F. $2t^2 - 5^t$ H. $3k^6 + k^4 - 4$

G. $\dfrac{1}{n^2 + 3n + 2}$ I. $7a^{3.5} + 5a^2 + 1$

8. **SHORT RESPONSE** An Australian Rocket Frog jumps off the ground. The path of its jump can be modeled by $y = -\dfrac{1}{6}x^2 + x$, where x is horizontal distance (in feet) and y is vertical distance (in feet).

Part A Find the horizontal distance at which the frog reaches its maximum height.

Distance _____ feet

Part B Find the maximum height of the jump.

Maximum height _____ feet

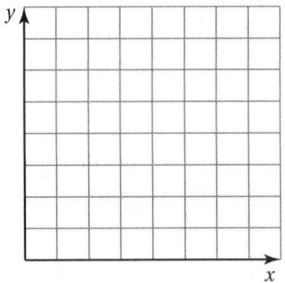

Part C Graph the equation of the jump.

Chapter 8 Standards Assessment Item Analysis

1.
   A. The student misapplies the Quotient of Powers Property.
   B. Correct answer
   C. The student misapplies the Power of Powers Property or the Quotient of Powers Property.
   D. The student misapplies the Product of Powers Property.

2.
   F. Correct answer
   G. The student thinks the terms of a quadratic function have constant first differences instead of constant second differences.
   H. The student thinks the terms of a quadratic function have a common ratio instead of constant second differences.
   I. The student thinks the terms of a quadratic function have constant third differences instead of constant second differences.

3. Correct answer: 9

   Common error: The student gives the growth factor 3 instead of the square of the growth factor.

4.
   A. The student gets the sign wrong for the vertical translation.
   B. The student gets the sign wrong for the horizontal translation.
   C. Correct answer
   D. The student gets the signs wrong for the vertical and horizontal translations.

5.
   F. The student only finds the y-intercepts of the lines in the graph.
   G. The student only finds the slopes of the lines in the graph.
   H. Correct answer
   I. The student correctly finds the equation for one of the lines, $y = 2x + 3$, but gets the other equation wrong.

6.
   A. Correct answer
   B. The student misapplies the rule for factoring the difference of two squares.
   C. The student incorrectly uses the rule for factoring perfect square trinomials.
   D. The student incorrectly uses the rule for factoring perfect square trinomials.

7.
   F. The student forgets that polynomials cannot have variable exponents.
   G. The student forgets that polynomials cannot have variables in the denominator.
   H. Correct answer
   I. The student forgets that polynomials must have whole number exponents.

Chapter 8 Standards Assessment Item Analysis (continued)

8. **2 points** The student demonstrates a thorough understanding of finding the vertex of and graphing a function of the form $y = ax^2 + bx + c$. In part A, the student should answer 3 feet. In part B, the student should answer 1.5 feet. In part C, the student should accurately draw a smooth curve through the points $(0, 0)$, $(3, 1.5)$, and $(6, 0)$.

   1 point The student's work demonstrates limited understanding of finding the vertex of and graphing a function of the form $y = ax^2 + bx + c$. Either Part A and B are incorrect due to a small error in the vertex formula or the graph is incorrectly drawn.

   0 points The student provides no response, a completely incorrect or incomprehensible response, or a response that demonstrates insufficient understanding of finding the vertex of and graphing a function of the form $y = ax^2 + bx + c$.

Chapter 8 Alternative Assessment

1. Use the table to complete the following questions.

Time, t	1	2	3	4
Distance, d	3	6	11	18

   a. Plot the points in a coordinate plane.

   b. Tell whether the data values represent a *linear*, an *exponential*, or a *quadratic* function. Explain.

   c. Write a function that models the data.

   d. When $t = 6$, what is the value of d?

   e. When $d = 83$, what is the value of t?

   f. Describe a real-life situation that can be represented by the table of values.

2. Graph each function. Then compare it to the graph of $y = x^2$. Find the vertex and axis of symmetry. Determine whether the graph opens up or down, and whether the graph is wider or narrower than $y = x^2$. Describe the domain and range of the function. State the minimum or maximum value.

   a. $y = 2x^2$

   b. $y = -3x^2$

   c. $y = \frac{1}{3}x^2$

   d. $y = -\frac{1}{2}x^2$

   e. $y = x^2 + 5$

   f. $y = (x + 3)^2 - 2$

   g. $y = -2x^2 + 4x$

   h. $y = x^2 + 2x - 3$

Name _____ Date _____

Chapter 8 Alternative Assessment Rubric

Score	Conceptual Understanding	Mathematical Skills	Work Habits
4	Shows complete understanding of: • graphing and writing quadratic functions • identifying the vertex, axis of symmetry, domain, range, and minimum or maximum value	In Exercise 1, correctly classifies function as quadratic and graphs the function correctly. Correctly graphs each function in Exercise 2 and correctly identifies each characteristic of the function.	Answers all parts of both problems. All equations and graphs are written or drawn carefully and systematically. Work is very neat and well organized.
3	Shows nearly complete understanding of: • graphing and writing quadratic functions • identifying the vertex, axis of symmetry, domain, range, and minimum or maximum value	In Exercise 1, correctly classifies function as quadratic and graphs the function correctly. Correctly graphs most of the functions in Exercise 2 and correctly identifies most characteristics of each function.	Answers several parts of both problems. Most equations and graphs are written or drawn carefully and systematically. Work is neat and organized.
2	Shows some understanding of: • graphing and writing quadratic functions • identifying the vertex, axis of symmetry, domain, range, and minimum or maximum value	In Exercise 1, incorrectly classifies function as quadratic and does not graph the function correctly. Correctly graphs some of the functions in Exercise 2 and correctly identifies some characteristics of each function.	Answers some parts of both problems. Equations and graphs are written or drawn carelessly. Work is not very neat or organized.
1	Shows little understanding of: • graphing and writing quadratic functions • identifying the vertex, axis of symmetry, domain, range, and minimum or maximum value	In Exercise 1, incorrectly classifies function as quadratic and does not graph the function correctly. Incorrectly graphs the functions in Exercise 2 and does not identify the characteristics of each function.	Does not attempt any part of either problem. No equations or graphs are written or drawn. Work is sloppy and disorganized.

Chapter 9 Quiz
For use after Section 9.3

Solve the equation by graphing. Check your solution(s).

1. $x^2 + 2x - 8 = 0$
2. $x^2 + 6x = -9$

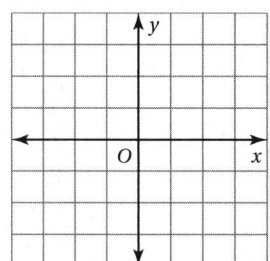 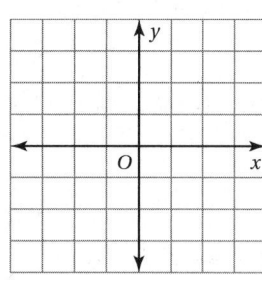

Solve the equation using square roots.

3. $2x^2 + 3 = 3$
4. $5x^2 - 20 = 0$
5. $3x^2 + 48 = 0$
6. $-2x^2 + 98 = 0$
7. $(x - 2)^2 = 9$
8. $\frac{1}{3}(x + 4)^2 = 0$

Complete the square for the expression. Then factor the trinomial.

9. $x^2 + 2x$
10. $x^2 - 6x$

Solve the equation by completing the square.

11. $x^2 - 4x = 10$
12. $x^2 - 10x = 11$
13. $2x^2 + 8x + 18 = 42$
14. $x^2 + 3x = -\frac{5}{4}$

15. A toy car launches off a ramp from a height of 6 feet with an upward velocity of 10 feet per second. The function $h = -16t^2 + 10t + 6$ gives the height h (in feet) of the car after t seconds. After how many seconds does the car land on the ground?

16. The area of the square is 50 square inches. Find the value of x. Round your answer to the nearest hundredth.

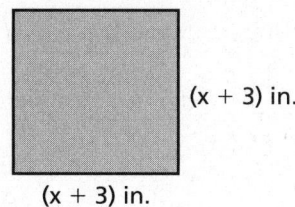

$(x + 3)$ in.

Answers

1. See left.
2. See left.
3. _____
4. _____
5. _____
6. _____
7. _____
8. _____
9. _____
10. _____
11. _____
12. _____
13. _____
14. _____
15. _____
16. _____

Name _____   Date _____

Chapter 9 Quiz
For use after Section 9.5

Solve the equation using the quadratic formula.

1. $2x^2 + 5x - 3 = 0$
2. $3x^2 - x - 10 = 0$
3. $x^2 + 6x + 12 = 0$
4. $3x^2 + 4x - 2 = 0$

Use the discriminate to determine the number of real solutions of the equation.

5. $x^2 + 7x + 13 = 0$
6. $2x^2 - 8x + 8 = 0$

Solve the equation using any method. Explain your choice of method.

7. $x^2 + 8x - 9 = 0$
8. $2x^2 - 5x - 3 = 0$

9. $2x^2 + 6x - 36 = 0$
10. $x^2 + 5x = 0$

Solve the system.

11. $y = x^2 - 36$
    $y = 9x$
12. $y - 3 = x^2$
    $y = 2$
13. $y = x^2 - 4x + 5$
    $y = x - 1$
14. $y = x^2 - 7x + 10$
    $y = -5x + 10$

15. The amount of money a store earns during each hour one day can be modeled by the function $y = -8x^2 + 64x - 60$, where y is in dollars and x is the time in hours from when the store opens. During which hour will the store earn $68?

16. The customers c for two new boutiques can be modeled by the following equations, where x is the number of days since the boutiques opened.

    $c = -x^2 + 6x + 7$ Boutique 1
    $c = 2x + 11$ Boutique 2

    When is the number of customers for each boutique the same?

Answers

1. _____
2. _____
3. _____
4. _____
5. _____
6. _____
7. _____
   See left.
8. _____
   See left.
9. _____
   See left.
10. _____
    See left.
11. _____
12. _____
13. _____
14. _____
15. _____
16. _____

Name_____ Date_____

Chapter 9 Test A

Solve the equation by graphing.

1. $x^2 + 4x - 5 = 0$
2. $-x^2 + 2x - 1 = 0$

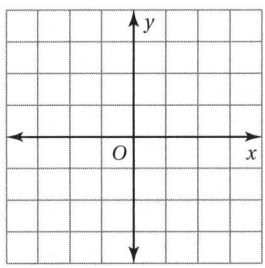
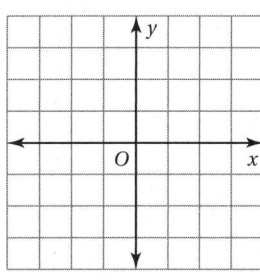

Solve the equation using square roots.

3. $x^2 - 16 = -16$
4. $4x^2 + 9 = 0$
5. $3x^2 - 12 = 0$
6. $(x + 2)^2 = 36$

7. The area of the rug is 27 square feet. Find the dimensions of the rug.

$(x - 2)$ ft
$(x + 4)$ ft

Solve the equation by completing the square.

8. $x^2 + 2x = 8$
9. $x^2 - 6x + 7 = 0$
10. $2x^2 + 8x - 24 = 0$
11. $x^2 + 5x = -\dfrac{9}{4}$

Solve the equation using the quadratic formula.

12. $x^2 - 3x - 9 = -5$
13. $2x^2 + 3x + 1 = 0$
14. $4x^2 - 49 = 0$
15. $3x^2 - 4x = -1$

Use the discriminate to determine the number of real solutions of the equation.

16. $x^2 - 3x + 5 = 0$
17. $-\dfrac{1}{2}x^2 - 4x + 1 = 0$
18. $2x^2 + 8x + 8 = 0$
19. $7x^2 - 2x + 1 = 0$

Answers

1. ____See left.____
2. ____See left.____
3. _____
4. _____
5. _____
6. _____
7. _____
8. _____
9. _____
10. _____
11. _____
12. _____
13. _____
14. _____
15. _____
16. _____
17. _____
18. _____
19. _____

Chapter 9 Test A (continued)

Solve the equation using any method. Explain your choice of method.

20. $x^2 - 6x + 4 = 0$

21. $x^2 + 7x - 30 = 0$

22. $2x^2 + 15x - 8 = 0$

23. $8x^2 - 64 = 0$

Solve the system.

24. $y = x - 4$
    $y = x^2 - 7x - 13$

25. $y = x^2$
    $y = 7x + 18$

26. $y = -3x$
    $y = x^2 - 9x + 20$

27. $y = \dfrac{3}{2}x - 1$
    $y = x^2 - \dfrac{1}{2}x$

Answers

20. _____See left._____
21. _____See left._____
22. _____See left._____
23. _____See left._____
24. _____
25. _____
26. _____
27. _____
28. _____
29. _____
30. _____
    _____

28. Find the value of b that makes $x^2 + bx + 36$ a perfect square trinomial.

29. A batter hits a baseball 4 feet above the ground with an upward velocity of 63 feet per second. The function $h(t) = -16t^2 + 63t + 4$ gives the height h (in feet) of the baseball after t seconds. How long is the ball in the air if no one catches it?

30. The volume of the loading space of a moving truck is about 432 cubic feet. Write and solve a quadratic equation to find the length of the loading space of the moving truck.

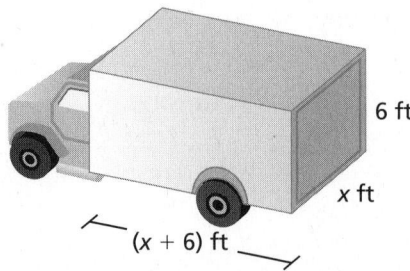

Name_____ Date_____

Chapter 9 Test B

Solve the equation by graphing.

1. $x^2 - 9 = 0$
2. $-x^2 + 2x + 3 = 0$

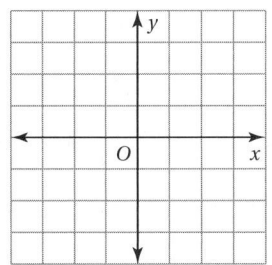
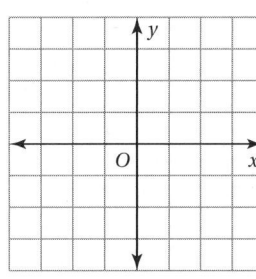

Solve the equation using square roots.

3. $-2x^2 - 18 = 0$
4. $3x^2 = 9$
5. $9(x-1)^2 = 16$
6. $4x^2 - 100 = 0$

7. The area of the rug is 49π square feet. Find the value of x.

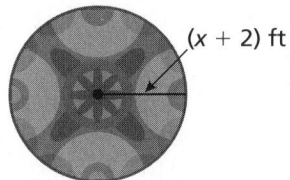

Solve the equation by completing the square.

8. $x^2 - 8x = 33$
9. $x^2 + 12x + 25 = 0$
10. $-2x^2 + 4x - 7 = 0$
11. $x^2 - 5x = \dfrac{11}{4}$

Solve the equation using the quadratic formula.

12. $x^2 - 2x - 10 = 14$
13. $9x^2 - 6x + 1 = 0$
14. $2x^2 + 5x = 3$
15. $2x^2 - 5 = 4x$

Use the discriminate to determine the number of real solutions of the equation.

16. $x^2 - 7x + 6 = 0$
17. $4x^2 - 24x + 36 = 0$
18. $-3x^2 - 1 = 6x$
19. $\dfrac{1}{2}x^2 + 3x = -8$

Answers

1. _See left._
2. _See left._
3. _____
4. _____
5. _____
6. _____
7. _____
8. _____
9. _____
10. _____
11. _____
12. _____
13. _____
14. _____
15. _____
16. _____
17. _____
18. _____
19. _____

Chapter 9 Test B (continued)

Solve the equation using any method. Explain your choice of method.

20. $x^2 - 10x - 24 = 0$

21. $16x^2 - 1 = 0$

22. $3x^2 - 2x = 5$

23. $6x^2 + 2x = -x$

Solve the system.

24. $y = -3x^2$
    $y = 14x + 8$

25. $y = -x + 5$
    $y = -x^2 + 2x + 1$

26. $y = -x$
    $y = x^2 + 4x - 14$

27. $y = x^2 - 2x + 5$
    $y = -x + 1$

Answers

20. _See left._
21. _See left._
22. _See left._
23. _See left._
24. _____
25. _____
26. _____
27. _____
28. _____
29. _____
30. _____

28. Find the value of b that makes $x^2 + bx + \dfrac{9}{4}$ a perfect square trinomial.

29. A batter hits a baseball 4 feet above the ground with an upward velocity of 49 feet per second. The function $h(t) = -16t^2 + 49t + 4$ gives the height h (in feet) of the baseball after t seconds. How long is the ball in the air if a player on the other team catches the ball 7 feet above the ground?

30. The volume of the loading space of a moving truck is about 768 cubic feet. Write and solve a quadratic equation to find the length of the loading space of the moving truck.

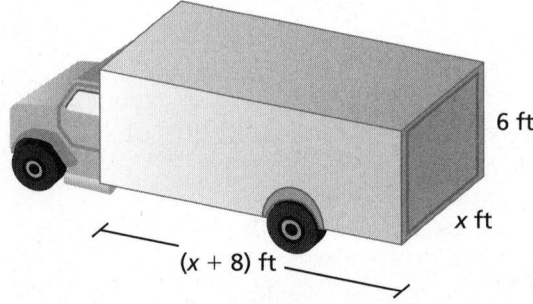

Chapter 9 Standards Assessment

1. Which of the following quadratic equations has exactly one real solution?

   A. $x^2 - 2x + 7 = 0$

   B. $x^2 - 3x - 10 = 0$

   C. $2x^2 + 4x - 2 = 0$

   D. $2x^2 - 8x + 8 = 0$

2. **GRIDDED RESPONSE** The focus of the parabola formed by $y = kx^2$ is $\left(0, \frac{1}{6}\right)$. What is the value of k?

3. The graph shows the height of a plant y, measured in inches, after x weeks. Which linear function relates y to x?

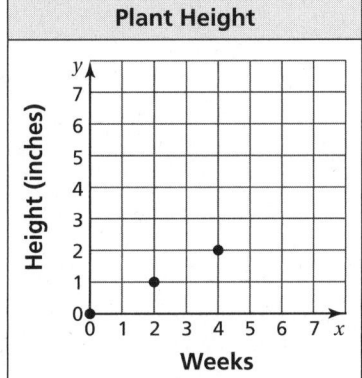

   F. $y = x - 1$

   G. $y = \frac{1}{2}x$

   H. $y = x - 2$

   I. $y = 2x$

4. The steps Dan took to solve an inequality are shown below. What should Dan change in order to correctly solve the inequality?

   $$3x + 2 > x - 10$$
   $$2x > -12$$
   $$x < -6$$

   A. Nothing; he solved the inequality correctly.

   B. The constants should combine to equal -8.

   C. The $<$ should be switched to a $>$ in the third line.

   D. The $>$ should be switched to a $<$ in the second line.

Chapter 9 Standards Assessment (continued)

5. The total area covered by four identical sidewalk squares is 36 square feet. What is the side length of one sidewalk square?

  F. 2 feet   H. 6 feet

  G. 3 feet   I. 9 feet

6. What type of equation is graphed?

  A. Linear

  B. Quadratic

  C. Exponential

  D. None of the above

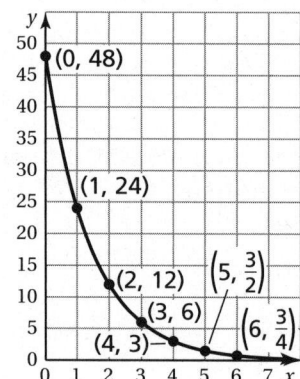

7. What are the binomial factors of $x^2 + 5x - 84$?

  F. $(x - 7)$ and $(x + 12)$ H. $(x - 8)$ and $(x + 13)$

  G. $(x + 7)$ and $(x - 12)$ I. $(x + 8)$ and $(x - 13)$

8. Which expression is equivalent to $\dfrac{3 - \sqrt{45}}{6}$?

  A. -2 C. $2 - \sqrt{5}$

  B. $\dfrac{1 - 3\sqrt{5}}{2}$ D. $\dfrac{1 - \sqrt{5}}{2}$

9. EXTENDED RESPONSE An amusement park is made up of a ride park and a water park. The attendances y (in hundreds) can be modeled by the equations below, where x is the number of days since the park opened for summer.

$$y = -x^2 + 32x + 9 \quad \text{Ride park}$$
$$y = 8x + 32 \quad \text{Water park}$$

  Part A When does the ride park have an attendance of 240 hundred people? Write what method you use to solve the equation and explain why you choose that method.

  Part B When do the two parks have the same attendance? Show all work necessary to justify your answer.

Chapter 9 Standards Assessment Item Analysis

1.
   A. The student thinks a negative discriminant implies exactly one real solution.
   B. The student thinks a perfect square discriminant implies exactly one real solution.
   C. The student gets the sign of $-4ac$ wrong when calculating the discriminant $b^2 - 4ac$.
   D. Correct answer

2. Correct answer: 1.5 or $\frac{3}{2}$

   Common error: The student sets $\frac{1}{4}k = \frac{1}{6}$ instead of $\frac{1}{4k} = \frac{1}{6}$ and answers $\frac{2}{3}$.

3.
   F. The student misreads the graph and picks an equation that passes through the point $(2, 1)$.
   G. Correct answer
   H. The student misreads the graph and picks an equation that passes through the point $(4, 2)$.
   I. The student inverts the slope.

4.
   A. The student incorrectly reverses the inequality when a negative number is added to both sides.
   B. The student adds 2 to the right-hand side instead of subtracting.
   C. Correct answer
   D. The student incorrectly reverses the inequality when both sides are divided by a positive number.

5.
   F. The student sets $x^2 = 4$ instead of $4x^2 = 36$.
   G. Correct answer
   H. The student sets $x^2 = 36$ instead of $4x^2 = 36$.
   I. The student solves for x^2 instead of for x.

6.
   A. The student thinks the terms of a linear function differ by a constant factor rather than a constant amount.
   B. The student thinks the terms of a quadratic function differ by a constant factor rather than have constant second differences.
   C. Correct answer
   D. The student doesn't know the properties of an exponential function or fails to notice them.

Chapter 9 Standards Assessment Item Analysis (continued)

7. **F.** Correct answer
   G. The student gets the sign of the *x*-coefficient wrong.
   H. The student checks if the factors produce the correct *x*-coefficient but doesn't check if they produce the same constant term.
   I. The student checks if the factors produce the correct *x*-coefficient but gets the sign wrong and doesn't check if they produce the same constant term.

8. **A.** The student incorrectly simplifies $\sqrt{45}$ as 15.
   B. The student incorrectly simplifies $\sqrt{45}$ as $9\sqrt{5}$ or forgets to divide $3\sqrt{5}$ by 3.
   C. The student switches the 3 and the 6 when simplifying.
   D. Correct answer

9. **4 points** The student demonstrates a thorough understanding of choosing a solution method for solving quadratic equations and solving systems of linear and quadratic equations. In Part A, the student writes the equation $-x^2 + 32x + 9 = 240$, solves it by factoring, completing the square, or using the quadratic formula, and gives a reasonable explanation for why he/she chose his/her method. The solutions are the 11th and 21st days. In Part B, the student clearly shows the steps of solving the system using either substitution or elimination and gets answers of the 1st and 23rd days.

   3 points The student demonstrates an essential but less than thorough understanding of choosing a solution method for solving quadratic equations and solving systems of linear and quadratic equations. The equations are set up correctly and the proper steps for the solution method are followed, but the student may have made a small arithmetic error in solving for *x*.

   2 points The student demonstrates a partial understanding of choosing a solution method for solving quadratic equations and solving systems of linear and quadratic equations. The student's work and explanations demonstrate a lack of essential understanding. For example, the student gives a good explanation for why he/she chose his/her solution method in Part A, but gets the process wrong.

   1 point The student demonstrates a limited understanding of setting up and solving systems of equations. The student's response is incomplete and exhibits many flaws. For example, the initial equations in Part A might be incorrect and the subsequent graphing inconsistent with these equations.

   0 points The student provided no response, a completely incorrect or incomprehensible response, or a response that demonstrates insufficient understanding of choosing a solution method for solving quadratic equations and solving systems of linear and quadratic equations.

Chapter 9 Alternative Assessment

1. Consider the equation $2x^2 + 9x + 4 = 0$.

   a. Use the discriminant to find the number of real solutions of the equation.

   b. Solve the equation by graphing.

   c. Solve the equation using two other methods. Show all of your work.

   d. Explain which method you would prefer to solve the equation.

2. Consider the equation $x^2 - 4x - 14 = 0$.

   a. Use the discriminant to find the number of real solutions of the equation.

   b. Solve the equation using any two methods. Show all of your work.

   c. Explain which method you would prefer to solve the equation.

3. Consider the equation $-x^2 + 6x - 9 = 0$.

   a. Use the discriminant to find the number of real solutions of the equation.

   b. Solve the equation by graphing.

   c. Solve the equation using two other methods. Show all of your work.

   d. Explain which method you would prefer to solve the equation.

4. Consider the equation $2x^2 + 18 = 0$.

   a. Use the discriminant to find the number of real solutions of the equation.

   b. Solve the equation using any two methods. Show all of your work.

   c. Explain which method you would prefer to solve the equation.

5. Explain the relationship between the discriminant, x-intercepts, and solutions of a quadratic equation.

Name_____ Date_____

Chapter 9 Alternative Assessment Rubric

Score	Conceptual Understanding	Mathematical Skills	Work Habits
4	Shows complete understanding of: • solving quadratic equations • finding the relationship between the discriminant, the x-intercept, and the solutions of a quadratic equation	Finds the discriminant of each equation and solves all equations correctly. Correctly explains the relationship between the discriminant, x-intercept, and solutions of a quadratic equation.	Answers all parts of each problem. All equations and graphs are written or drawn carefully and systematically. Work is very neat and well organized.
3	Shows nearly complete understanding of: • solving quadratic equations • finding the relationship between the discriminant, the x-intercept, and the solutions of a quadratic equation	Finds most discriminants of the equations and solves most equations correctly. Correctly explains the relationship between the discriminant, x-intercept, and solutions of a quadratic equation.	Answers several parts of each problem. Most equations and graphs are written or drawn carefully and systematically. Work is neat and organized.
2	Shows some understanding of: • solving quadratic equations • finding the relationship between the discriminant, the x-intercept, and the solutions of a quadratic equation	Finds some discriminants of the equations and solves some equations correctly. Incorrectly explains the relationship between the discriminant, x-intercept, and solutions of a quadratic equation.	Answers some parts of each problem. Equations and graphs are written or drawn carelessly. Work is not very neat or organized.
1	Shows little understanding of: • solving quadratic equations • finding the relationship between the discriminant, the x-intercept, and the solutions of a quadratic equation	Does not find the discriminant of each equation. Incorrectly solves each equation. Does not explain the relationship between the discriminant, x-intercept, and solutions of a quadratic equation.	Does not attempt any part of either problem. No equations or graphs are written or drawn. Work is sloppy and disorganized.

Name_____ Date_____

Chapter 10 Quiz
For use after Section 10.2

Find the domain of the function.

1. $y = \sqrt{x} - 7$

2. $y = -3\sqrt{x+2}$

Graph the function. Describe the domain and range. Compare the graph to the graph of $y = \sqrt{x}$.

3. $y = \sqrt{x} + 3$

4. $y = -\sqrt{x-2}$

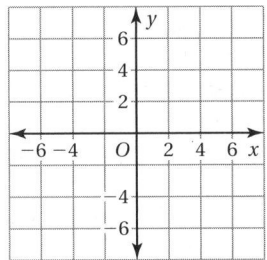
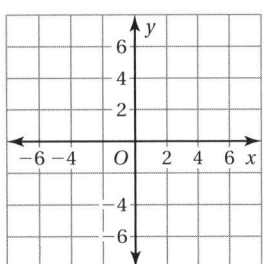

Solve the equation.

5. $\sqrt{x} = 4$

6. $\sqrt{x} - 2 = 10$

7. $-2\sqrt{x} + 3 = 1$

8. $\sqrt{x-1} - 5 = 1$

9. $\sqrt{4x-1} = 2x$

10. $\sqrt{3x-2} = \sqrt{2x+5}$

Find the value of x.

11. Perimeter = 14 in.

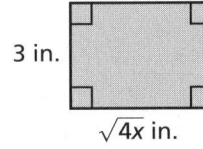

12. Area = 20 m²

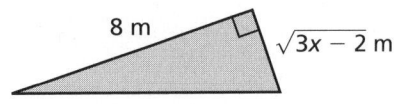

13. The time t (in seconds) it takes you to swing forward and back on a rope is given by the function $t = 2\pi\sqrt{\dfrac{r}{32}}$, where r is the rope length (in feet). It takes 3 seconds to swing forward and back. How long is the rope? Use 3.14 for π. Round your answer to the nearest tenth.

Answers

1. _____
2. _____
3. ___See left.___
_____
_____
_____
_____
_____
_____
4. ___See left.___
_____
_____
_____
_____
_____
_____
5. _____
6. _____
7. _____
8. _____
9. _____
10. _____
11. _____
12. _____
13. _____

Name _____ Date _____

Chapter 10 Quiz
For use after Section 10.4

Find the missing length of the triangle.

1.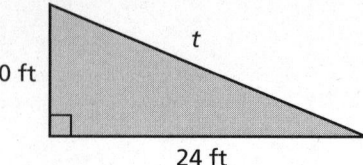

2.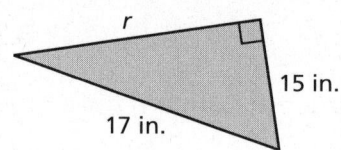

3.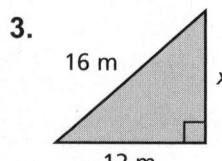

4.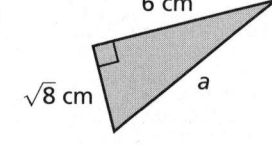

Tell whether the triangle with the given side lengths is a right triangle.

5.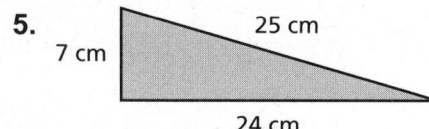

6.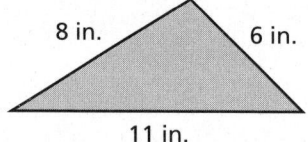

Find the distance between the two points.

7. $(0, 0), (3, 4)$

8. $(2, -3), (7, -15)$

9. $(-5, -1), (0, 2)$

10. $(-1, 4), (-3, -2)$

11. You ride your bike on the path shown. Does the path form a right triangle?

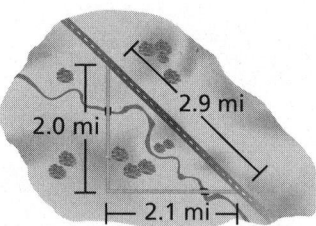

Use the figure to answer Exercises 12-15. Each grid represents 10 feet. Round your answers to the nearest tenth.

12. How far is your locker from the cafeteria?

13. How far is math class from the cafeteria?

14. How far is your locker from math class?

15. Does the path form a right triangle?

Answers

1. _____
2. _____
3. _____
4. _____
5. _____
6. _____
7. _____
8. _____
9. _____
10. _____
11. _____
12. _____
13. _____
14. _____
15. _____

Name_____ Date_____

Chapter 10 Test A

Graph the function. Describe the domain and range. Compare the graph to the graph of $y = \sqrt{x}$.

1. $y = \sqrt{x-1}$

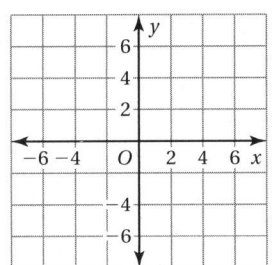

2. $y = -\sqrt{x} + 1$

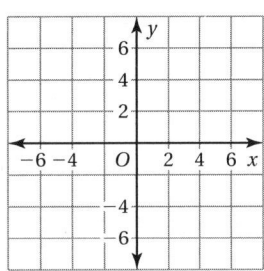

Solve the equation.

3. $\sqrt{x} = 9$

4. $\sqrt{x-3} = 6$

5. $\sqrt{x-2} + 3 = 7$

6. $\sqrt{2x+29} = x - 3$

7. $\sqrt{x} = \sqrt{2x-4}$

8. $\sqrt{3x-1} - \sqrt{x+5} = 0$

9. The perimeter of the square is 12 feet. Find the value of x.

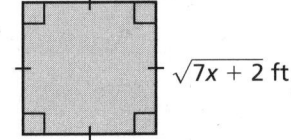

Find the missing length of the triangle.

10.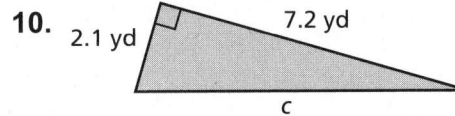

11.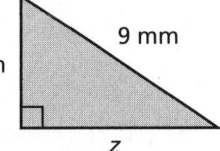

Tell whether the triangle with the given side lengths is a right triangle.

12.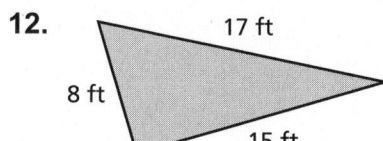

13. 9 in., 7 in., 13 in.

Answers

1. ___See left.___

_____
_____
_____
_____
_____
_____

2. ___See left.___

_____
_____
_____
_____
_____
_____

3. _____
4. _____
5. _____
6. _____
7. _____
8. _____
9. _____
10. _____
11. _____
12. _____
13. _____

Name _____ Date _____

Chapter 10 Test A (continued)

Find the distance between the two points.

14. $(0, 0), (8, -15)$ 15. $(1, -2), (4, 2)$

16. $(-5, -1), (2, 3)$ 17. $(4, -1), (-2, -9)$

18. $(1, 5), (4, 2)$ 19. $(0, -6), (-1, 3)$

20. Consider the graph of $y = \sqrt{x}$. If you translate the graph 3 units to the left and 2 units down, what is the range of the translation?

21. The formula $V = \sqrt{PR}$ relates the voltage V (in volts), power P (in watts), and resistance R (in ohms) of an electrical circuit. What is the resistance of a 100-watt stereo system on a 120-volt circuit?

22. You are walking your dog with a 5-foot leash. What is the farthest distance your dog can be from you? Round your answer to the nearest tenth.

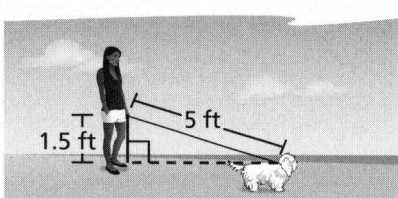

23. A bicycle frame is made up of two triangles. The larger triangle is a right triangle.

   a. Find the value of x. Round your answer to the nearest tenth.

   b. Does the smaller triangle form a right triangle? Explain.

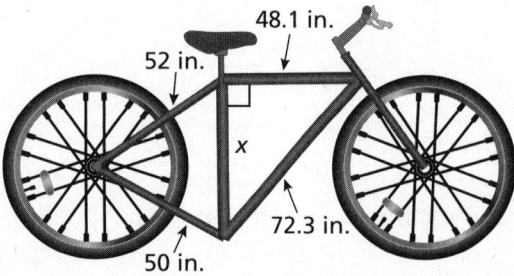

Answers

14. _____
15. _____
16. _____
17. _____
18. _____
19. _____
20. _____
21. _____
22. _____
23. a. _____

    b. _____

    _____

    _____

    _____

Name_____ Date_____

Chapter 10 Test B

Graph the function. Describe the domain and range. Compare the graph to the graph of $y = \sqrt{x}$.

1. $y = \sqrt{3 - x}$

2. $y = \sqrt{x - 2} + 3$

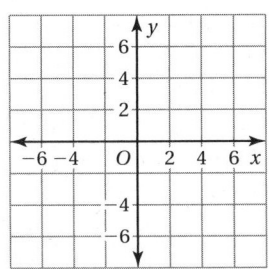 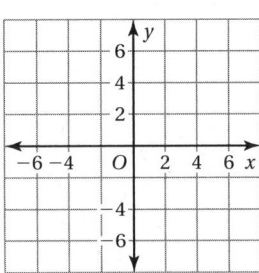

Answers

1. _____See left._____

2. _____See left._____

Solve the equation.

3. $\sqrt{x} = 5$

4. $\sqrt{x} + 2 = 10$

5. $-\sqrt{x - 7} + 3 = -1$

6. $\sqrt{2x + 5} + 1 = 2x$

7. $\sqrt{x + 3} = \sqrt{4x}$

8. $\sqrt{x - 1} - \sqrt{3x - 7} = 0$

9. The area of the rectangle is 56 square centimeters. Find the value of x.

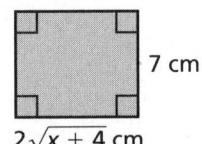

Find the missing length of the triangle.

10.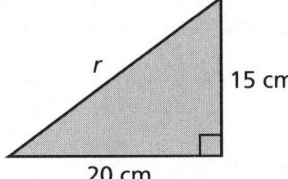

11.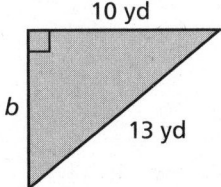

Tell whether the triangle with the given side lengths is a right triangle.

12.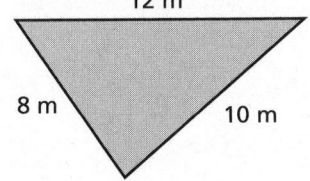

13.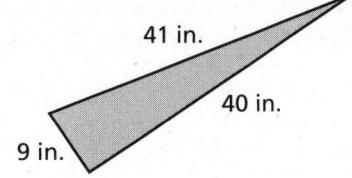

3. _____

4. _____

5. _____

6. _____

7. _____

8. _____

9. _____

10. _____

11. _____

12. _____

13. _____

Name _____ Date _____

Chapter 10 Test B (continued)

Find the distance between the two points.

Answers

14. $(0, 0), (-5, -12)$ 15. $(7, 0), (-2, 3)$

16. $(-2, 2), (-3, 1)$ 17. $(3, 2), (-3, -6)$

18. $(2, -1), (-3, 2)$ 19. $(-2, 4), (-4, 0)$

20. Consider the graph of $y = \sqrt{x}$. If you reflect the graph in the x-axis, and translate the graph 3 units to the left, what is the range of the translation?

21. The formula $V = \sqrt{PR}$ relates the voltage V (in volts), power P (in watts), and resistance R (in ohms) of an electrical circuit. What is the resistance of a 3600-watt air conditioner on a 240-volt circuit?

22. You are flying two kites. Which kite is closer to you? Explain.

14. _____

15. _____

16. _____

17. _____

18. _____

19. _____

20. _____

21. _____

22. _____

_____

_____

_____

23. a. _____

b. _____

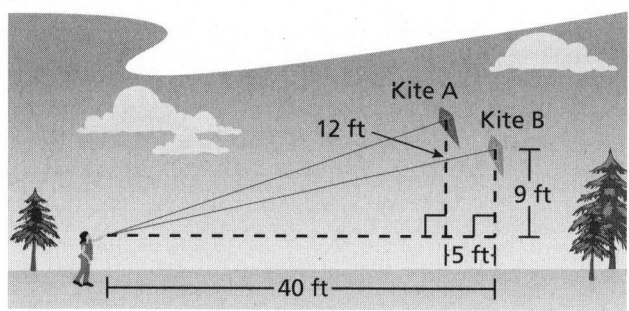

23. A truss, which is used in construction, is formed by triangles.

   a. Find the value of x.

   b. Find the entire length of the truss.

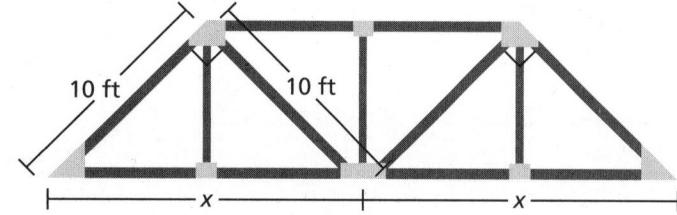

118 Big Ideas Math Algebra 1
Assessment Book

Chapter 10 Standards Assessment

1. Mary graphed $y = \sqrt{x-1} - 1$. Her work is shown on the right. What should Mary change in order to graph the equation correctly?

   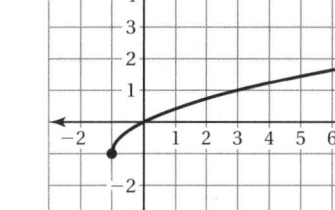

   A. Nothing; she graphed the equation correctly.

   B. Shift the graph 2 units right.

   C. Shift the graph 2 units up.

   D. Shift the graph 2 units up and 2 units right.

2. In the system of equations below, a and b are constants.

   $$y = ax + 5$$
   $$y = bx + 5$$

   What can you conclude?

   F. The system of equations has no solution.

   G. The system of equations has exactly one solution.

   H. The system of equations has at least one solution.

   I. The system of equations has infinitely many solutions.

3. **GRIDDED RESPONSE** Find the side length x of the triangle. Round to the nearest tenth of a foot.

   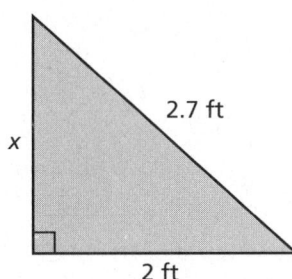

4. Which polynomial can be multiplied by $x + 3$ to produce $3x^3 + 13x^2 + 12x$?

   A. $3x + 4$ C. $3x^2 + 4x$

   B. $4x + 3$ D. $4x^2 + 3x$

Chapter 10 Standards Assessment (continued)

5. A parabola has a focus at $(0, -6)$ and vertex at the origin. What is an equation of this parabola?

   F. $y = -6x^2$

   G. $y = -\dfrac{3}{2}x^2$

   H. $y = -\dfrac{1}{6}x^2$

   I. $y = -\dfrac{1}{24}x^2$

6. Which of the following relations is not a function?

   A. $(-2, 3), (-1, 4), (0, 5),$ $(-1, 6), (-2, 7)$

   C.
Input	0	1	2	3	4
Output	-3	-1	1	3	5

   B. Input, x Output, y
   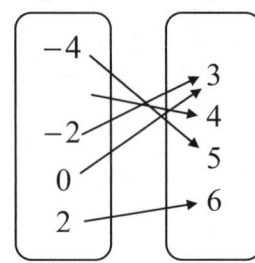

   D.
   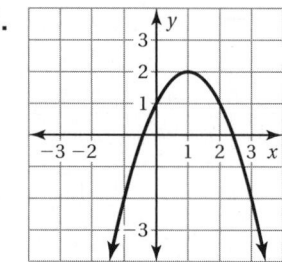

7. Antonio is solving the equation $x^2 - 8x = 5$. Which number should he add to each side to solve the equation by completing the square?

   F. -64

   G. -16

   H. 16

   I. 64

8. **SHORT RESPONSE** The groundwork for a new building has been laid. One corner of the building is shown on the right. Does the corner form a right angle? Explain your reasoning.

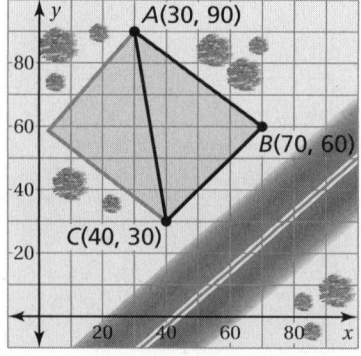

Chapter 10 Standards Assessment Item Analysis

1. **A.** The student thinks $y = \sqrt{x-1} - 1$ is a translation 1 unit left and 1 unit down of $y = \sqrt{x}$.
   B. Correct answer
   C. The student thinks $y = \sqrt{x-1} - 1$ is a translation 1 unit left and 1 unit up of $y = \sqrt{x}$.
   D. The student thinks $y = \sqrt{x-1} - 1$ is a translation 1 unit right and 1 unit up of $y = \sqrt{x}$.

2. **F.** The student thinks the lines are parallel and that there is no solution.
   G. The student realizes that $(0, 5)$ is a solution but doesn't consider that the lines could be the same.
   H. Correct answer
   I. The student incorrectly assumes the lines are the same.

3. Correct answer: 1.8
   Common error: The student computes $\sqrt{2.7^2 + 2^2} \approx 3.4$ instead of $\sqrt{2.7^2 - 2^2}$.

4. **A.** The student gets the coefficients of the factor correct, but factors the product as if the degree of each term is one less than it is.
   B. The student switches the coefficients of the factor and factors the product as if the degree of each term is one less than it is.
   C. Correct answer
   D. The student switches the coefficients of the factor.

5. **F.** The student lets the a in $y = ax^2$ equal the y-coordinate of the focus instead of setting $-6 = \dfrac{1}{4a}$.
   G. The student sets $-6 = 4a$ instead of $-6 = \dfrac{1}{4a}$.
   H. The student sets $-6 = \dfrac{1}{a}$ instead of $-6 = \dfrac{1}{4a}$.
   I. Correct answer

6. **A.** Correct answer
   B. The student thinks every output must have exactly one input for a relation to be a function.
   C. The student does not understand the definition of a function.
   D. The student uses a horizontal line test instead of the vertical line test to determine if the relation is a function.

Chapter 10 Standards Assessment Item Analysis (continued)

7. **F.** The student subtracts the square of -8 instead of adding the square of -8 divided by 2.

   G. The student subtracts the square of -8 divided by 2 instead of adding it.

   H. Correct answer

   I. The student adds the square of -8 instead of adding the square of -8 divided by 2.

8. **2 points** The student demonstrates a thorough understanding of using the distance formula and the converse of the Pythagorean Theorem. The student explains that $AB^2 + BC^2 = 2500 + 1800 = 4300$, but $AC^2 = 3700$ so the points don't form a right triangle and the building corner does not form a right angle.

   1 point The student demonstrates a partial understanding of using the distance formula and the converse of the Pythagorean Theorem. For example, the distances are calculated correctly but the Pythagorean Theorem is applied incorrectly or vice-versa.

   0 points The student provides no response, a completely incorrect or incomprehensible response, or a response that demonstrates insufficient understanding of using the distance formula and the converse of the Pythagorean Theorem.

Name_____ Date _____

Chapter 10 Alternative Assessment

1. Consider the function $f(x) = \sqrt{x-2} + 5$.

   a. Sketch the graph of the function.

   b. Compare the graph to the graph of $y = \sqrt{x}$.

   c. Describe the domain and range.

   d. Find $f(11)$ and $f(1)$.

   e. Find $f(x) = 7$ and $f(x) = 4$.

2. Use the map to answer the following questions.

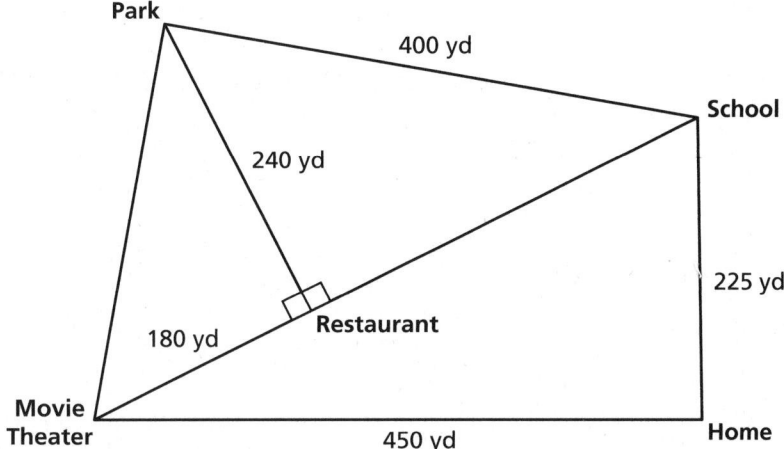

   a. Find the distance between the movie theater and the park.

   b. Find the distance between the restaurant and the school.

   c. Is the triangle formed by the park, movie theater, and school a right triangle? Explain.

   d. Is the triangle formed by the movie theater, school, and home a right triangle? Explain.

Name _____ Date _____

Chapter 10 Alternative Assessment Rubric

Score	Conceptual Understanding	Mathematical Skills	Work Habits
4	Shows complete understanding of: • graphing, evaluating, and solving square root functions • applying the Pythagorean Theorem	Correctly graphs, evaluates, and solves the function in Exercise 1. In Exercise 2, correctly uses the Pythagorean Theorem to find the missing distances, and correctly determines if the two larger triangles are right triangles.	Answers all parts of both problems. All equations and graphs are written or drawn carefully and systematically. Work is very neat and well organized.
3	Shows nearly complete understanding of: • graphing, evaluating, and solving square root functions • applying the Pythagorean Theorem	Graphs, evaluates, or solves the function in Exercise 1 with 1 or 2 errors. In Exercise 2, uses the Pythagorean Theorem to find one missing distance, and determines if the two larger triangles are right triangles with 1 or 2 errors.	Answers several parts of both problems. Most equations and graphs are written or drawn carefully and systematically. Work is neat and organized.
2	Shows some understanding of: • graphing, evaluating, and solving square root functions • applying the Pythagorean Theorem	Graphs, evaluates, or solves the function in Exercise 1 with a few errors. In Exercise 2, uses the Pythagorean Theorem to find the missing distances, and determines if the two larger triangles are right triangles with a few errors.	Answers some parts of both problems. Equations and graphs are written or drawn carelessly. Work is not very neat or organized.
1	Shows little understanding of: • graphing, evaluating, and solving square root functions • applying the Pythagorean Theorem	Incorrectly graphs, evaluates, and solves the function in Exercise 1. In Exercise 2, incorrectly uses the Pythagorean Theorem to find the missing distances, and incorrectly determines if the two larger triangles are right triangles.	Does not attempt any part of either problem. No equations or graphs are written or drawn. Work is sloppy and disorganized.

Name_____ Date_____

Chapter 11 Quiz
For use after Section 11.3

Tell whether *x* and *y* show *direct variation*, *inverse variation*, or *neither*.

1.
x	1	2	3	4
y	30	15	10	7.5

2.
x	1	2	3	4
y	−4	−8	−12	−16

3. The variable *y* varies directly with *x*. When $x = 8$, $y = 6$. Write and graph a direct variation equation that relates *x* and *y*.

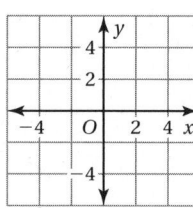

Identify the asymptotes of the graph of the function. Then describe the domain and range.

4. $y = \dfrac{3}{x-1} + 2$

5. $y = -\dfrac{1}{x} - 3$

Find the inverse of the function. Graph the inverse function.

6. $f(x) = 4x - 1$

7. $f(x) = x^2 + 2$, where $x \geq 0$

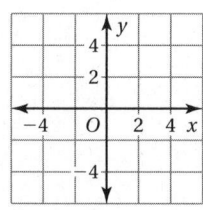
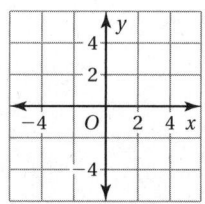

Simplify the rational expression, if possible. State the excluded value(s).

8. $\dfrac{3t^7}{15t^5}$

9. $\dfrac{18w^2}{2w^7}$

10. $\dfrac{a^2 - 9}{a^2 + 2a - 3}$

11. $\dfrac{b^2 + 8b + 15}{b^2 - b - 12}$

12. You have $120 to donate equally to several charities. Does this situation represent direct variation or inverse variation? Explain.

13. The side length of a square is represented by the expression $x + 1$. Write and simplify a rational expression for the ratio of the perimeter to the area.

Answers

1. _____
2. _____
3. _____
   See left.
4. _____
   _____
   _____
   _____
   _____
5. _____
   _____
   _____
   _____
6. _____
   See left.
7. _____
   See left.
8. _____
9. _____
10. _____
11. _____
12. _____
    _____
    _____
    _____
13. _____

Name _____ **Date** _____

Chapter 11 Quiz
For use after Section 11.7

Find the product or quotient.

1. $\dfrac{3z}{5} \cdot \dfrac{10z}{9z^3}$

2. $\dfrac{3m}{m+3} \cdot \dfrac{m^2 + 6m + 9}{6m^2}$

3. $\dfrac{x+6}{x-2} \div \dfrac{x^2 - 36}{x^2 - 3x + 2}$

4. $\dfrac{7b^3}{b^2 - 9b + 20} \div \dfrac{14b^4}{3b - 12}$

Find the quotient.

5. $(p^2 + 3p - 28) \div (p - 4)$

6. $(c^2 - 5c - 8) \div (c + 2)$

Find the sum or difference.

7. $\dfrac{y^2 - 4y}{y - 3} + \dfrac{3y - 6}{y - 3}$

8. $\dfrac{x - 3}{x^2 + 2x + 1} + \dfrac{2}{x + 1}$

9. $\dfrac{k + 1}{5k^2} - \dfrac{3}{10k}$

10. $\dfrac{m}{2m - 6} - \dfrac{m + 1}{m^2 - 9}$

Solve the equation. Check your solution.

11. $\dfrac{2}{x - 2} = \dfrac{6}{x - 8}$

12. $\dfrac{4}{x + 10} = \dfrac{x}{x - 2}$

13. $\dfrac{3}{c} - \dfrac{1}{2} = \dfrac{c - 3}{c}$

14. $\dfrac{3}{a - 3} + \dfrac{2}{a + 3} = \dfrac{5a + 3}{a^2 - 9}$

15. The cost for you and your friends to go to a movie is represented by $6x + 24$, where x is the number of tickets sold. Everybody who went to the movie, except 2 friends who forgot their money, shares the cost equally. Find $(6x + 24) \div (x - 2)$ to determine an expression for how much each person pays.

16. You have a bag of peanuts, chocolate, and raisins. Fifteen of the 36 pieces are peanuts. You add more peanuts and increase the ratio of peanuts to the total number of peanuts, chocolate, and raisins to 9 : 16. How many peanuts do you add?

Answers

1. _____
2. _____
3. _____
4. _____
5. _____
6. _____
7. _____
8. _____
9. _____
10. _____
11. _____
12. _____
13. _____
14. _____
15. _____
16. _____

Name_____ Date_____

Chapter 11 Test A

1. The variable y varies directly with x. When $x = 5$, $y = -30$. Write and graph a direct variation equation that relates x and y.

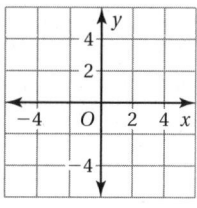

2. The variable y varies inversely with x. When $x = 4$, $y = \dfrac{1}{2}$. Write and graph an inverse variation equation that relates x and y.

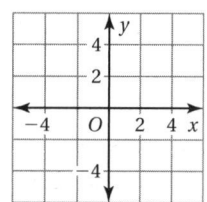

3. You can either buy 4 tickets for $22 or 6 tickets for $33. Determine whether the situation represents direct variation or inverse variation. Justify your answer.

Graph the function. Compare the graph to the graph of $y = \dfrac{1}{x}$.

4. $y = \dfrac{1}{x + 2}$

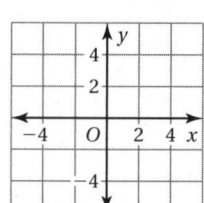

5. $y = -\dfrac{1}{x} + 4$

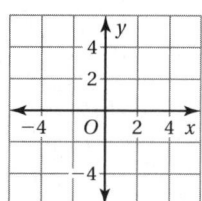

Find the inverse of the function. Graph the inverse function.

6. $f(x) = -2x + 4$

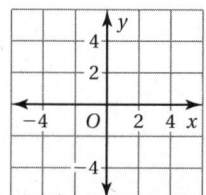

7. $f(x) = x^2 - 4$, where $x \geq 0$

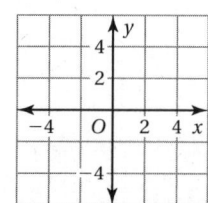

Answers

1. _____
   See left.

2. _____
   See left.

3. _____
   _____
   _____
   _____

4. _____ See left.
   _____
   _____
   _____

5. _____ See left.
   _____
   _____
   _____

6. _____
   See left.

7. _____
   See left.

Name _____ Date _____

Chapter 11 Test A (continued)

Simplify.

8. $\dfrac{21f^5}{7f^6}$

9. $\dfrac{4a^2 + 12a + 8}{2a^2 - 8}$

10. $\dfrac{u+2}{8u^2} \cdot \dfrac{4u^2}{u^2 + 5u + 6}$

11. $\dfrac{g-3}{5g^3} \div \dfrac{2g-6}{10g^2}$

12. $(q^2 + 3q - 18) \div (q - 3)$

13. $(e^2 - 4e + 7) \div (e - 3)$

14. $\dfrac{3x^2}{x+2} - \dfrac{x-4}{x+2}$

15. $\dfrac{c+4}{c^2-1} + \dfrac{2c}{c-1}$

Solve the equation. Check your solution.

16. $\dfrac{3}{x+8} = \dfrac{2}{x+14}$

17. $\dfrac{x}{x-1} = \dfrac{6}{x+1}$

18. $\dfrac{2}{m} + \dfrac{3}{5} = \dfrac{m-4}{m}$

19. $\dfrac{4}{y-2} - \dfrac{7}{y+3} = \dfrac{5y+2}{y^2+y-6}$

20. You download 10 songs on several occasions. The average time y (in songs per minute) is given by $y = \dfrac{10}{x}$, where x is the time (in minutes) it takes to download 10 songs. Graph the function. Make a conclusion from the graph.

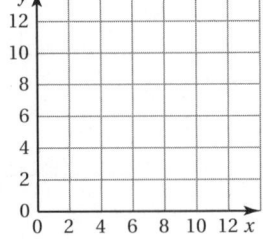

21. You can paint a room in 6 hours. Your friend can paint the same room in 10 hours. Working together, how much time does it take to paint the room?

Answers

8. _____

9. _____

10. _____

11. _____

12. _____

13. _____

14. _____

15. _____

16. _____

17. _____

18. _____

19. _____

20. __See left.__

_____

_____

_____

_____

21. _____

Name_____ Date_____

Chapter 11 Test B

Answers

1. The variable y varies directly with x. When $x = -3$, $y = 9$. Write and graph a direct variation equation that relates x and y.

1. _____
   See left.

2. The variable y varies inversely with x. When $x = 8$, $y = \dfrac{3}{2}$. Write and graph an inverse variation equation that relates x and y.

2. _____
   See left.

3. _____

3. You can either babysit 3 children at $4 per hour for each child or 4 children at $3 per hour for each child. Determine whether the situation represents direct variation or inverse variation. Justify your answer.

4. See left.

Graph the function. Compare the graph to the graph of $y = \dfrac{1}{x}$.

5. See left.

4. $y = \dfrac{1}{x - 1} + 5$

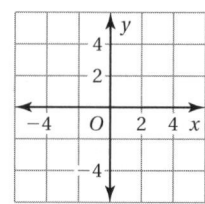

5. $y = -\dfrac{1}{x + 3} - 1$

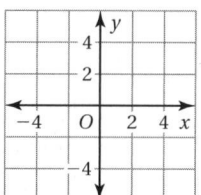

6. _____
   See left.

Find the inverse of the function. Graph the inverse function.

7. _____
   See left.

6. $f(x) = \dfrac{1}{2}x + 2$

7. $f(x) = -\dfrac{1}{2}x^2$, where $x \geq 0$

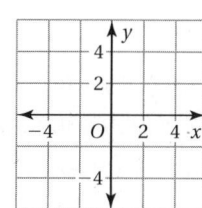

Copyright © Big Ideas Learning, LLC
All rights reserved.

Big Ideas Math Algebra 1
Assessment Book

Name _____ Date _____

Chapter 11 Test B (continued)

Simplify.

8. $\dfrac{10d^8}{25d^7}$

9. $\dfrac{4x^3 - 4x^2 + x}{6x^3 + 15x^2 - 6x}$

10. $\dfrac{w+3}{2w^2} \cdot \dfrac{8w^3}{3w+9}$

11. $\dfrac{n^2 - 25}{5n} \div \dfrac{n^2 - 10n + 25}{10n^2}$

12. $(6k^2 - 7k - 5) \div (2k + 1)$

13. $(2y^2 - 6) \div (y + 3)$

14. $\dfrac{x+1}{4x} + \dfrac{x-1}{3x}$

15. $\dfrac{7w+3}{w^2+4w-12} - \dfrac{2w}{w-2}$

Solve the equation. Check your solution

16. $\dfrac{6}{x-1} = \dfrac{9}{x+1}$

17. $\dfrac{x}{x+4} = \dfrac{10}{4x-2}$

18. $\dfrac{3}{d-4} + \dfrac{12}{d} = \dfrac{d+1}{d-4}$

19. $\dfrac{7}{x-5} - \dfrac{5}{x-2} = \dfrac{3x-1}{x^2-7x+10}$

20. The amount of money y (in dollars per hour) that an employee earns on a project varies inversely with the amount t of time spent on the project. An employee who works 5 hours earns $15 an hour on the project.

   a. Write an inverse variation equation that relates t and y.

   b. How much does the employee earn for completing the project?

   c. How many hours did the employee work when the amount earned is $10 an hour?

21. You can paint a room in 9 hours. Your friend can paint the same room in 12 hours. Working together, how much time does it take to paint the room?

Answers

8. _____
9. _____
10. _____
11. _____
12. _____
13. _____
14. _____
15. _____
16. _____
17. _____
18. _____
19. _____
20. a, _____
    b. _____
    c, _____
21. _____

Chapter 11 Standards Assessment

1. Which of the following expressions is not equivalent to a rational number?

   A. $-\dfrac{3}{4} + 1.2$

   B. $4.5 - \pi$

   C. $\sqrt{8} \cdot \sqrt{2}$

   D. $(2.4)(3.71)$

2. **GRIDDED RESPONSE** The square below has an area of $(n + 20)$ square feet. Find the value of n.

   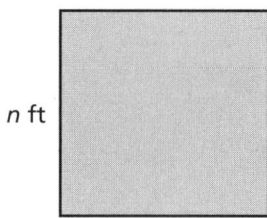

   n ft

3. The steps Robert took to divide $3x^2 - 8$ by $x - 2$ are shown below. What should Robert change in order to find the quotient correctly?

   $$\begin{array}{r} 3x + 6 \\ x - 2 \overline{\smash{\big)}\, 3x^2 + 0x - 8} \\ \underline{3x^2 - 6x} \\ 6x - 8 \\ \underline{6x - 12} \\ 4 \end{array}$$

   So, $(3x^2 - 8) \div (x - 2) = 3x + 6 + \dfrac{4}{x - 2}$.

   F. Nothing; he found the quotient correctly.

   G. The last term of the quotient should be 4.

   H. The x-term of the dividend should not be considered in the division process.

   I. The expression $6x - 8$ should be $-6x - 8$.

4. The variable y varies directly with x. When $x = -6$, $y = -15$. Which equation relates y and x?

   A. $y = -\dfrac{2}{5}x$

   B. $y = \dfrac{2}{5}x$

   C. $y = -\dfrac{5}{2}x$

   D. $y = \dfrac{5}{2}x$

Chapter 11 Standards Assessment (continued)

5. What is the equation of the parabola shown in the graph?

   F. $y = x^2 - 3$

   G. $y = 2x^2 - 3$

   H. $y = x^2 + 3$

   I. $y = 2x^2 + 3$

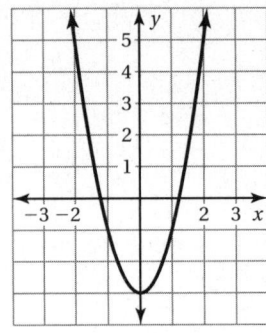

6. What is the solution to $x = 21x^2 + 7x$?

   A. $x = -\dfrac{1}{3}$

   B. $x = -\dfrac{2}{7}$

   C. $x = -\dfrac{1}{3}, 0$

   D. $x = -\dfrac{2}{7}, 0$

7. Which graph represents the inequality $x > -4$?

   F.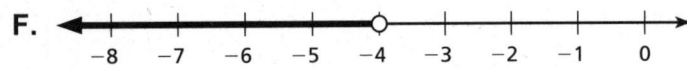

   G.

   H.

   I.

8. **EXTENDED RESPONSE** A competitor in a bicycle race rides the first 14 miles at a constant speed r (in miles per hour). It starts raining and she reduces her speed by three miles per hour and finishes the last six miles of the race at that pace.

   Part A Write a rational expression that represents the total time it takes her to complete the race. Show all work necessary to justify your answer.

   Part B It takes her an hour to complete the race. Find her initial speed r. Show all work necessary to justify your answer.

Chapter 11 Standards Assessment Item Analysis

1. A. The student assumes the sum of a decimal and a fraction is irrational.
    B. Correct answer
    C. The student assumes the product of two irrational numbers is irrational.
    D. The student assumes that because the product does not simplify to an "easy" fraction it is irrational.

2. Correct answer: 5
    Common error: The student gives the extraneous solution -4.

3. F. Correct answer
    G. The student forgets the remainder is equal to the final term of the long division process divided by the divisor, not the final term itself.
    H. The student excludes the x-term of the divisor because its coefficient is 0.
    I. The student subtracts -6 from 0 incorrectly.

4. A. The student thinks the coefficient of variation should be negative because x and y are negative, and confuses x and y.
    B. The student confuses x and y.
    C. The student thinks the coefficient of variation should be negative because x and y are negative.
    D. Correct answer

5. F. The student correctly translates $y = x^2$ vertically but doesn't realize the graph is too narrow to be $y = x^2 - 3$.
    G. Correct answer
    H. The student uses the wrong sign to translate $y = x^2$ vertically and does not realize the graph is too narrow to be $y = x^2 + 3$.
    I. The student uses the wrong sign to translate $y = x^2$ vertically.

6. A. The student sets the right-hand side of the equation equal to zero and divides out x but doesn't recognize $x = 0$ as a solution.
    B. The student divides out x but doesn't recognize $x = 0$ as a solution.
    C. The student sets the right-hand side of the equation equal to zero.
    D. Correct answer

7. F. The student confuses $<$ and $>$.
    G. The student confuses $<$ and $>$, and chooses the closed circle at -4 when it should be an open circle because the symbol is "greater than."
    H. Correct answer
    I. The student chooses the closed circle at -4 when it should be an open circle because the symbol is "greater than."

Chapter 11 Standards Assessment Item Analysis (continued)

8. **4 points** The student demonstrates a thorough understanding of writing and adding rational expressions and solving rational equations. Each part is answered correctly and clearly. For Part A, the student writes the terms $\frac{14}{r}$ and $\frac{6}{r-3}$ and adds them together getting a total time of $\frac{20r-42}{r^2-3r}$ hours. For Part B, the student sets $\frac{20r-42}{r^2-3r} = 1$ and answers $r = 21$ miles per hour. The student disregards the other root of the equation, $r = 2$, because it would imply a negative speed.

   3 points The student demonstrates an essential but less than thorough understanding of writing and adding rational expressions and solving rational equations. For example, the student gives $r = 2$ miles per hour as an answer for Part B or uses the right approach to solve the rational equation but makes a small error.

   2 points The student demonstrates a partial understanding of writing and adding rational expressions and solving rational equations. The student's work and explanations demonstrate a lack of essential understanding. For example, the student makes a mistake in finding the least common denominator in Part A.

   1 point The student demonstrates limited understanding of writing and adding rational expressions and solving rational equations. The student's response is incomplete and exhibits many flaws.

   0 points The student provides no response, a completely incorrect or incomprehensible response, or a response that demonstrates insufficient understanding of writing and adding rational expressions and solving rational equations.

Chapter 11 Alternative Assessment

1. Consider the function $f(x) = \dfrac{1}{x+4} - 3$.

   a. Sketch the graph of the function.

   b. Identify the asymptotes and describe the domain and range.

   c. Find the inverse of the function.

   d. Sketch the graph of the inverse function.

   e. Identify the asymptotes of the inverse function.

2. Use the expressions to complete the following.

   A. $\dfrac{x^7}{x^2 - x}$ B. $\dfrac{x^2 - 4x + 3}{x^5}$ C. $\dfrac{x^2 - 1}{x^4}$

   a. Which expression is not simplified? Simplify the expression. State any excluded values.

   b. Find the product of any two expressions. Explain your choice. State any excluded values.

   c. Find the quotient of any two expressions. Explain your choice. State any excluded values.

   d. Find the sum and difference of any two expressions. Explain your choice.

   e. Find the value of x that makes $B = \dfrac{1}{x} \cdot C$ true.

Name _____ Date _____

Chapter 11 Alternative Assessment Rubric

Score	Conceptual Understanding	Mathematical Skills	Work Habits
4	Shows complete understanding of: • graphing rational functions • finding the inverse of functions • simplifying, adding, subtracting, multiplying and dividing rational expressions	Correctly graphs the rational function and finds and graphs the inverse function. Correctly simplifies, adds, subtracts, multiplies, and divides all of the rational expressions and solves the rational equation.	Answers all parts of both problems. All equations and graphs are written or drawn carefully and systematically. Work is very neat and well organized.
3	Shows nearly complete understanding of: • graphing rational functions • finding the inverse of functions • simplifying, adding, subtracting, multiplying and dividing rational expressions	Graphs the rational function and finds and graphs the inverse function with 1 or 2 errors. Correctly simplifies, adds, subtracts, multiplies, and divides most of the rational expressions and solves the rational equation.	Answers several parts of both problems. Most equations and graphs are written or drawn carefully and systematically. Work is neat and organized.
2	Shows some understanding of: • graphing rational functions • finding the inverse of functions • simplifying, adding, subtracting, multiplying and dividing rational expressions	Graphs the rational function and finds and graphs the inverse function with a few errors. Correctly simplifies, adds, subtracts, multiplies, and divides some of the rational expressions and solves the rational equation.	Answers some parts of both problems. Equations and graphs are written or drawn carelessly. Work is not very neat or organized.
1	Shows little understanding of: • graphing rational functions • finding the inverse of functions • simplifying, adding, subtracting, multiplying and dividing rational expressions	Incorrectly graphs the rational function and incorrectly finds the inverse function. Incorrectly simplifies, adds, subtracts, multiplies, and divides the rational expressions and incorrectly solves the rational equation.	Does not attempt any part of either problem. No equations or graphs are written or drawn. Work is sloppy and disorganized.

Name_____ Date _____

Chapter 12 Quiz
For use after Section 12.4

Find the mean, median, and mode of the data.

1.

Pairs of Shoes Owned		
4	5	8
6	10	2
9	12	7

2.

Numbers of Siblings		
2	1	0
4	5	3
0	1	2

3. The times (in seconds) for two sprinters are shown. Find the mean, range, and standard deviation of the data sets. Compare the data sets.

Joseph		
36	35	37
39	34	41

Daniel		
40	37	36
38	37	40

4. You record the ages of people attending a movie. Display the data in a histogram. Describe the shape of the distribution.

Age	Frequency
8–10	9
11–13	11
14–16	7
17–19	4
20–22	2

People Attending a Movie

(Frequency vs. Age histogram)

5. The numbers of points scored in each basketball game are 48, 36, 61, 64, 39, 41, 43, and 39.

  a. Make a box-and-whisker plot for the data.

  b. Find and interpret the interquartile range of the data.

  c. Describe the distribution of the data.

Answers

1. _____
   _____
   _____

2. _____
   _____
   _____

3. _____
   _____
   _____
   _____
   _____
   _____

4. ___See left.___

5. a. ___See left.___
   b. _____
   _____
   _____
   c. _____
   _____
   _____

Name _____ Date _____

Chapter 12 Quiz
For use after Section 12.8

1. The scatter plot shows the numbers of male teachers in a school district from 2004 to 2010.

   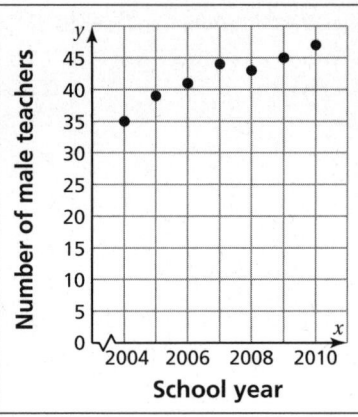

   a. In what school year did the school district have 41 male teachers?

   b. How many male teachers did the district have in the 2010 school year?

   c. Describe the relationship shown by the data.

2. The table shows the numbers of members at a gym each month.

Month	1	2	3	4	5	6	7
Members	20	34	40	45	50	52	60

   a. Make a scatter plot of the data and draw a line of fit.

   b. Write an equation of the line that fits the data.

   c. Interpret the slope of the line.

   d. Predict how many members the gym will have in one year.

   Gym Membership
   (graph with y-axis: Members, x-axis: Month)

3. You randomly survey students about their involvement in school sports and the school music program. The two-way table shows the results. Find and interpret the marginal frequencies for the survey.

		Sports		
		Involved	Not involved	Total
Music Program	Involved	62	34	
	Not involved	41	27	
	Total			

Answers

1. a. _____

   b. _____

   c. _____

2. a. __See left.__

   b. _____

   c. _____

3. __See left.__

   _____

   _____

   _____

138 Big Ideas Math Algebra 1
Assessment Book

Chapter 12 Test A

Find the mean, median, and mode of the data.

1.
Books Read		
5	3	3
3	2	5
2	3	1

2.
Distances Traveled (in miles)		
31	27	40
54	56	54
37	30	40

3. The sandwich prices for two restaurants are shown. Find the mean, range, and standard deviation of the data sets. Then compare the data sets.

Sam's Sandwiches		
$5	$6	$8
$6	$7	$7

Hugo's Hoagies		
$4	$5	$7
$8	$5	$7

4. The times (in minutes) spent fishing before getting a bite: 12, 25, 28, 20, 18, 16, 22, 30.

   a. Make a box-and-whisker plot of the data.

   b. Find the range and interquartile range.

   c. Describe the distribution of the data.

Choose an appropriate data display for the situation. Explain your reasoning.

5. the profits of a company over a year

6. a person's income based on age

7. number of wins for a baseball team each year

Answers

1. _____

2. _____

3. _____

4. a. _See left._

   b. _____

   c. _____

5. _____

6. _____

7. _____

Chapter 12 Test A (continued)

8. The table shows the total inches of rain that had fallen after each hour.

Hour	0	1	2	3	4	5
Inches of rain	0	0.5	1.1	1.8	2.4	3.0

a. Make a scatter plot of the data and draw a line of fit.

b. Write an equation of the line that fits the data.

c. Interpret the slope of the line.

d. If it continues to rain at a similar rate, predict how much rain will have fallen after 8 hours.

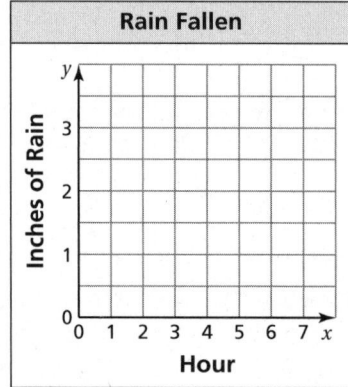

Answers

8. a. See left.
b. _____
c. _____
d. _____

9. a. See left.
b. _____
_____
_____
c. _____
_____

9. You randomly survey students about whether they ate or skipped lunch and breakfast. The results of the survey are shown in the two-way table.

a. How many students in the survey skipped breakfast but ate lunch?

b. How many of the students in the survey ate lunch?

c. How many students were surveyed?

		Breakfast		
		Ate	Skipped	Total
Lunch	Ate	40	12	
	Skipped	8	0	
	Total			

d. Find and interpret the marginal frequencies for the survey.

e. What percent of students skipped breakfast but ate lunch?

Name_____ Date_____

Chapter 12 Test B

Find the mean, median, and mode of the data.

1.

Dogs Boarded				
17	19	9	14	18
19	18	12	15	19

2.

Temperatures (°F)				
−4	−1	5	3	2
6	−10	−5	−1	0

3. The scores on a test from two classes are shown. Find the mean, range, and standard deviation of the data sets. Then compare the data sets.

Class A		
85	87	95
99	93	86
71	82	94

Class B		
75	77	93
86	83	95
83	92	81

4. The point spreads on 12 football games for a season are:

1, 3, 14, 9, 7, 3, 6, 27, 3, 13, 8, and 17.

a. Make a box-and-whisker plot for the data.

b. Find the range and interquartile range.

c. Describe the distribution of the data.

d. Identify the outlier. Explain how removing the outlier will affect the plot.

Choose an appropriate data display for the situation. Explain your reasoning.

5. percent of student athletes in each sport

6. the numbers and types of different species of fish at an aquarium

Answers

1. _____
   _____
   _____

2. _____
   _____
   _____

3. _____
   _____
   _____
   _____
   _____

4. a. _See left._
   b. _____
   _____
   c. _____
   d. _____
   _____
   _____

5. _____

6. _____

Chapter 12 Test B (continued)

7. The table shows the numbers of birds observed at a feeder each week.

Week	1	2	3	4	5	6
Birds	46	40	39	35	30	27

a. Make a scatter plot of the data and draw a line of fit.

b. Write an equation of the line that fits the data.

c. Interpret the slope of the line.

d. Estimate how many birds were at the feeder 3 weeks before week 1.

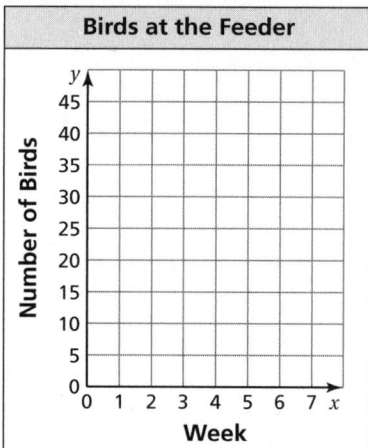
Birds at the Feeder

Answers

7. a. ___See left.___

b. _____

c. _____

d. _____

8. a. ___See left.___

b. _____

_____

_____

c. _____

_____

8. You randomly survey students in your class about whether they visited an amusement park in the summer. The results are shown.

Visited an Amusement Park
Boys: 38
Girls: 29

Did Not Visit an Amusement Park
Boys: 17
Girls: 25

a. Make a two-way table that includes the marginal frequencies.

		Visited an Amusement Park		
		Yes	No	Total
Student	Boys			
	Girls			
	Total			

b. Interpret the marginal frequencies for the survey.

c. For each gender, what percent of the students visited an amusement park?

Name_____ Date_____

Chapter 12 Standards Assessment

1. The histogram shows the credit scores of 100 adults.

   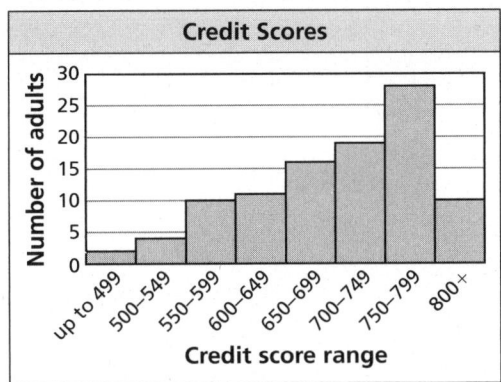

   Which measure of central tendency best represents these data?

   A. Mean   **C.** Mode

   B. Median   **D.** Standard deviation

2. What is the range of the equation below?

   $$y = -2|x - 3| + 1$$

   F. $y \le 1$ **H.** $y \le 2$

   G. $y \ge 1$ **I.** $y \ge 2$

3. **GRIDDED RESPONSE** The distance between the two points graphed below is 17. Find x.

   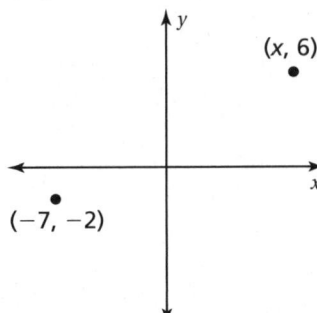

4. Which expression is equivalent to $\dfrac{2q^2 - 2q - 24}{q^3 + 7q^2 + 12q}$?

   A. $\dfrac{2}{q}$ **C.** $\dfrac{2q - 8}{q^2 + 3q}$

   B. $\dfrac{q - 4}{q + 4}$ **D.** $\dfrac{2q - 8}{q^2 + 4q}$

Chapter 12 Standards Assessment (continued)

5. The graph shows the linear relationship between the number of game tickets x and the total cost y to attend a school fair. What is the slope of the line?

    F. 0.8
    G. 1
    H. 1.25
    I. 2

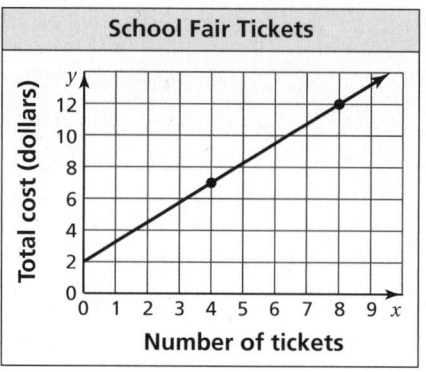

6. The **median** age of the 15 players on a professional basketball team is 27. Suppose a 30-year-old player is traded from this team in exchange for a 24-year-old player from another team. Which statement is true?

    A. The median age must decrease.
    B. The median age must increase.
    C. The median age could remain 27.
    D. The range must decrease.

7. At what values of x do the following two equations intersect?

    $$y = 2x^2 + x - 5$$
    $$y = 7x - 8$$

    F. $\dfrac{3 \pm \sqrt{3}}{2}$
    G. $\dfrac{3 \pm 2\sqrt{3}}{2}$
    H. $\dfrac{-4 \pm \sqrt{10}}{2}$
    I. $\dfrac{3 \pm \sqrt{15}}{2}$

8. What is the equation for the vertical asymptote of the graph of $y = \dfrac{4}{x + 3} + 2$?

    A. $x = -5$
    B. $x = -3$
    C. $x = 2$
    D. $x = 4$

9. **SHORT RESPONSE** The table shows the annual expenses of a company.

Category	Salary	Rent	Insurance	Utilities	Other
Annual Expenses (thousands of dollars)	450	84	36	24	50

You are asked how this data could be displayed effectively. Make **two** suggestions, and explain why you think each is a good choice.

Chapter 12 Standards Assessment Item Analysis

1.
   A. The student is thinking of the best measure of central tendency to describe a symmetric distribution instead of a skewed distribution.
   B. Correct answer
   C. The student realizes mean is not the best measure for a skewed distribution but incorrectly guesses mode.
   D. The student forgets that standard deviation is not a measure of central tendency.

2.
   F. Correct answer
   G. The student doesn't consider the effect of the negative sign in front of 2.
   H. The student thinks the multiplier –2 doubles the lower bound of the range.
   I. The student thinks the multiplier –2 doubles the lower bound of the range and doesn't consider the effect of the negative sign in front of 2.

3. Correct answer: 8

   Common error: The student substitutes the given coordinates into the formula $d = \sqrt{(x_2 - x_1)^2 + (y_2 - y_1)^2}$ incorrectly.

4.
   A. The student incorrectly divides out $q - 4$ and $q + 4$ as if they were common factors.
   B. The student incorrectly divides out 2 from the numerator and q from the denominator.
   C. The student incorrectly factors $2q^2 - 2q - 24$ as $2(q - 3)(q + 4)$ and divides out the $q + 4$ factors.
   D. Correct answer

5.
   F. The student reverses the roles of x and y in the slope formula.
   G. The student miscounts squares when calculating slope.
   H. Correct answer
   I. The student picks the y-intercept instead of the slope.

6.
   A. The student is thinking of the mean, not the median.
   B. The student is confused by the problem and thinks that the 30-year-old player is added to the team.
   C. Correct answer
   D. The student assumes that replacing an older player with a younger player decreases the range.

Chapter 12 Standards Assessment Item Analysis (continued)

7. **F.** Correct answer

   G. The student incorrectly factored $\sqrt{12}$ as $4\sqrt{3}$ when simplifying the result of the quadratic formula.

   H. The student added $7x$ to both sides instead of subtracting when solving the system by substitution.

   I. The student used $\sqrt{b^2 + 4ac}$ in the quadratic formula instead of $\sqrt{b^2 - 4ac}$.

8. **A.** The student finds the root of the function.

   B. Correct answer

   C. The student finds the horizontal asymptote $(y = 2)$ and then confuses the variables.

   D. The student thinks the numerator of the term $\dfrac{4}{x + 3}$ is the vertical asymptote.

9. **2 points** The student chooses two effective ways to display the data in the table, and provides a clear, persuasive explanation of why each is a good choice.

   1 point The student provides only one effective way to display the data (with thorough explanation) or provides insufficient explanations for two appropriate methods of display.

   0 points The student provides no response, a completely incorrect or incomprehensible response, or a response that demonstrates insufficient understanding of data displays.

Name_____ Date_____

 **Chapter 12 Alternative Assessment**

1. Sam gathered some data as he was doing homework this week.

Data Set 1	
Prices of Boxes of 500 Paper Clips	
Number of Boxes	Price per Box
0–5	$1.00
6–10	$0.80
11–25	$0.65
25 +	$0.55

Data Set 2	
Hours of Daylight at 50° North Latitude	
On the 15th of:	Hours of Daylight
January	8.5
February	10.1
March	11.8
April	13.8
May	17.1
June	16.4
July	15.6
August	14.6
September	12.7
October	10.6
November	9.1
December	8.1

Data Set 3	
Cars Rented in One Week at the Local Car Rental Agency	
Car Type	Number Rented
Economy	180
Compact	140
Standard	100
Luxury	60
SUV	80

a. Explain what kind of graph Sam should use for each data set and describe how he can make that graph.

b. For Data Set 2, find the mean, median, mode, range, and make a box-and-whisker plot. Then use your measures and plot to make a statement about this data set.

2. Joelle has gathered data about cats for a report. She wants to see if there is a line of fit that she can use to make predictions about other cats.

Cat	Length	Weight	Speed
Cheetah	7 ft	110 lb	70 mi/h
House Cat	2 ft	11 lb	30 mi/h
Lion	10 ft	330 lb	50 mi/h
Tiger	13 ft	550 lb	50 mi/h

a. Analyze this data using what you have learned in this chapter. Find a line of fit if you can, even if you have to remove an outlier.

b. Write three statements about the cats that are supported by the data.

Name _____ Date _____

Chapter 12 Alternative Assessment Rubric

Score	Conceptual Understanding	Mathematical Skills	Work Habits
4	Shows complete understanding of: • choosing graphs for data • finding and using measures of central tendency and the range of a data set • analyzing data	Matched all data sets and graphs correctly. Made the box-and-whisker plot using all correctly calculated measures. Analyzed the data in several ways, found a line of best fit, and wrote three statements.	Answers both parts of both questions. All calculations are done carefully. All work is neat and well organized.
3	Shows nearly complete understanding of: • choosing graphs for data • finding and using measures of central tendency and the range of a data set • analyzing data	Matched two data sets and graphs correctly. Made the box-and-whisker plot, but incorrectly calculated some measures. Analyzed the data in two ways, attempted to find a line of best fit, and wrote two statements.	Answers part of both questions. Most of the calculations are done carefully. Most of the work is neat and well organized.
2	Shows some understanding of: • choosing graphs for data • finding and using measures of central tendency and the range of a data set • analyzing data	Matched one data set and graph correctly. Made the box-and-whisker plot, but incorrectly calculated all measures. Analyzed the data in one way, did not attempt to find a line of best fit, and wrote one statement.	Answers only one part of one question. Some calculations are done carefully. Some work is neat and well organized.
1	Shows little understanding of: • choosing graphs for data • finding and using measures of central tendency and the range of a data set • analyzing data	Did not match any data sets and graphs correctly. Did not make the box-and-whisker plot. Did not analyze the data, but wrote one or two statements.	Answers few parts of the questions. No calculations are done carefully. All work is sloppy and disorganized.

Test 1: End-of-Course Test

Solve.

1. $x - 7 = -13$

2. $15 - 3c = 3$

3. One cell phone plan charges $20 per month plus $0.15 per minute used. A second cell phone plan charges $35 per month plus $0.10 per minute used. Write and solve an equation to find the number of minutes you must talk to have the same cost for both calling plans.

4. a. Write the formula for the area of a triangle.
   b. Solve the formula for h.
   c. The area of a triangle is 36 square inches. Use the new formula to find the height of the triangle in inches and in centimeters.

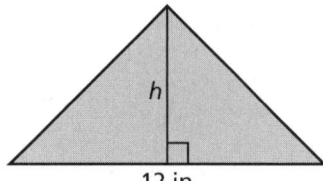

Find the slope and *y*-intercept of the graph of the linear equation. Then sketch its graph.

5. $y = 3x - 2$

6. $2x + 4y = 6$

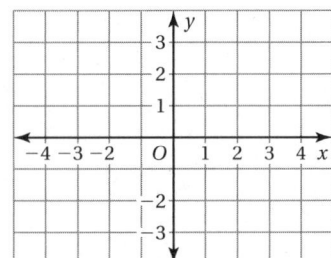 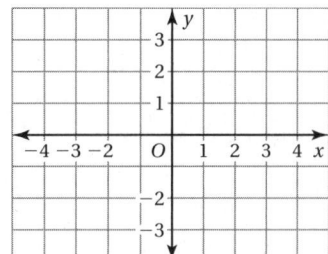

7. The equation $5x + 2y = 20$ represents the cost for a family to attend a play where x is the number of adults and y is the number of children. Find the intercepts and interpret the meaning of each one.

Solve the system.

8. $y = 3x + 4$
   $y - x = 2$

9. $y - 4x = 3$
   $2y = 8x + 5$

10. $y = x^2 - 9$
    $y = x + 3$

11. It costs $0.05 to send a text message and $0.10 to send a picture on your cell phone. You spend $4 and send five more text messages than pictures. How many text messages x and pictures y did you send?

Write an equation of the line in slope-intercept form.

12. the line passing through $(-1, 3)$ and $(-4, 5)$

13. the line with slope -2.5 and passing through $(2, 1.5)$

Answers

1. _____
2. _____
3. _____
   _____
4. a. _____
   b. _____
   c. _____
   _____
5. _____
   _____ See left.
6. _____
   _____
   _____ See left.
7. _____
   _____
   _____
   _____
8. _____
9. _____
10. _____
11. _____
    _____
12. _____
13. _____

Test 1 End-of-Course Test (continued)

14. Recall that 0°C = 32°F and 100°C = 212°F.

   a. Using x for degrees Celsius and y for degrees Fahrenheit, find an equation of the line passing through $(0, 32)$ and $(100, 212)$.

   b. What is the slope of the line? Explain what the slope means in terms of degrees Celsius and degrees Fahrenheit.

   c. What is the y-intercept of the line? Explain what the y-intercept means in terms of degrees Celsius and degrees Fahrenheit.

Determine the domain and range of the function.

15.
x	−2	−1	0	1	2	3
y	−3	−1	1	3	5	7

16.
x	−3	−2	−1	0	1	2
y	3	4	2	0	2	4

17. The table shows the cost y (in dollars) of x cold drinks.

Drinks, x	0	2	4	6
Cost, y	0	3	6	9

   a. Graph the data.

   b. Is the domain discrete or continuous?

   c. Write a linear function that relates y to x.

   d. How much does it cost to buy three drinks?

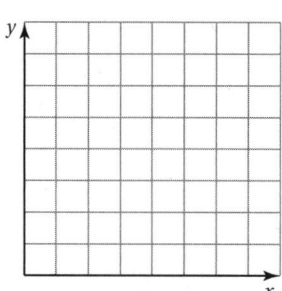

Simplify the expression.

18. $\dfrac{21x^3 y}{14x^5 y^2}$

19. $\left(\dfrac{3a^3 b^2}{2ab}\right)^{-2}$

20. $\sqrt[3]{125}$

21. $64^{2/3}$

22. $(3v^2 - 7v + 5) - (2v^2 + 3v - 7)$

23. $(2z + 3)^2$

24. $\dfrac{3x}{x^2 + 4x + 4} \div \dfrac{12x^3}{x^2 - x - 6}$

25. $\dfrac{3}{g} + \dfrac{g}{g - 4}$

Factor the polynomial.

26. $s^2 - 14s + 48$

27. $b^3 - 49b$

Answers

14. a. _____
    b. _____
    _____
    c. _____
    _____

15. _____
    _____

16. _____
    _____

17. a. See left.
    b. _____
    c. _____
    d. _____

18. _____
19. _____
20. _____
21. _____
22. _____
23. _____
24. _____
25. _____
26. _____
27. _____

Test 1 End-of-Course Test (continued)

28. A ladder is placed against the side of a house. The top of the ladder is 12 feet above the ground. The base of the ladder is 5 feet away from the house. Find the length of the ladder.

29. The data represent the weights (in pounds) of dogs at a dog show.

    8, 30, 37, 42, 50, 45, 35, 32, 40, 40, 55, 90

    a. Find the mean, median, and mode.

    b. Make a box-and-whisker plot for the data.

30. The table shows the number of years of college education and hourly earnings (in dollars) for several people.

Number of years, x	0	1	3	5	6
Hourly earnings, y	6	8	15	25	30

    a. Make a scatter plot of the data.

    b. Draw a line of fit.

    c. Write an equation for the line of fit.

    d. Predict the hourly earnings for a person with four years of college education.

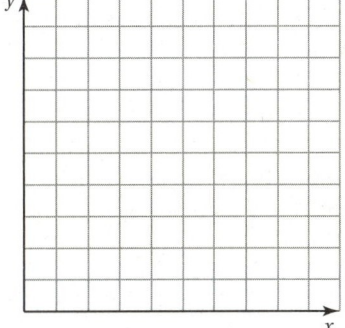

Choose an appropriate display for the situation. Explain your reasoning.

31. The percent of students with 0, 1, 2, or more than 2 siblings

32. The average movie theater ticket price over the last ten years

Write the word sentence as an inequality.

33. 3 less than a number t is at most 7.

34. A number m multiplied by 4 is greater than 12.

Answers

28. _____
29. a. _____
    _____
    _____
    b. _See left._
30. a. _See left._
    b. _See left._
    c. _____
    d. _____
31. _____
    _____
    _____
    _____
    _____
32. _____
    _____
    _____
    _____
33. _____
34. _____

Name _____ Date _____

Test 1 End-of-Course Test (continued)

Solve the inequality. Graph the solution.

35. $x + 4 > -6$

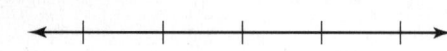

36. $3x - 2 \leq 7$

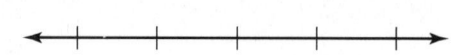

Answers

35. _____
    See left.

36. _____
    See left.

37. If you spend at least $50 (including shipping) at an online store, you receive a $10 gift card. You want to purchase CDs that cost $12 each. If shipping costs $5, write and solve an inequality to find the number of CDs you must buy to receive the gift card.

37. _____
    _____

Solve the equation. Check your solution(s).

38. $2^{x-3} = 8$

39. $a^2 + 5a - 14 = 0$

40. $r^2 + 4r - 16 = 0$

41. $2z^2 - 5z + 1 = 0$

42. $\sqrt{x + 4} - 9 = 2$

43. $\dfrac{3}{x + 5} = \dfrac{x}{x + 1}$

38. _____

39. _____

40. _____

41. _____

42. _____

43. _____

Graph the function. Describe the domain and range.

44. $y = (x - 3)^2 - 4$

45. $y = -\dfrac{1}{3}x^2$

44. See left.

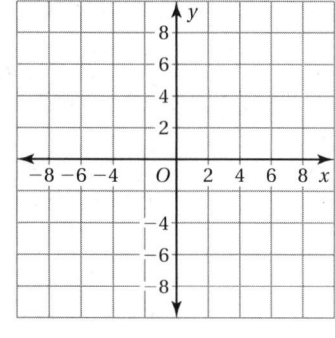   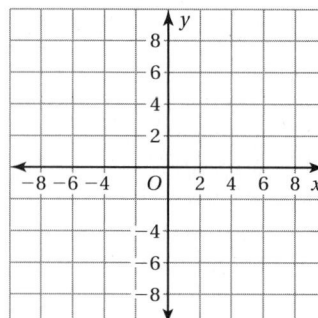

45. See left.

46. $y = -\sqrt{x + 1}$

47. $y = \dfrac{1}{x - 4} + 2$

46. See left.

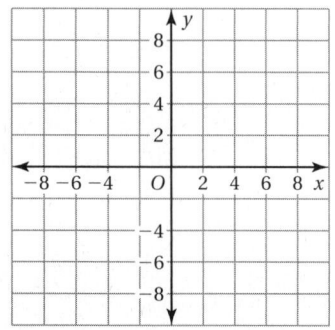   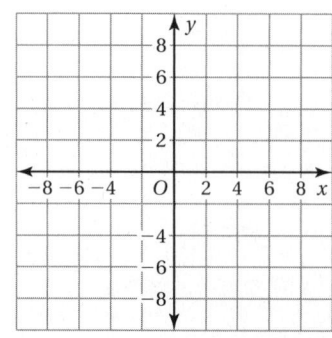

47. See left.

Name _____ Date _____

Test 2 — End-of-Course Test

Solve.

1. $r - 3.4 = -5.8$
2. $-1 - 2c = 4$

3. One cell phone plan charges $17.50 per month plus $0.17 per minute used. A second cell phone plan charges $32 per month plus $0.07 per minute used. Write and solve an equation to find the number of minutes you must talk to have the same cost for both calling plans.

4. a. Write the formula for the area of a triangle. Then solve for h.

   b. The area of a triangle is 14.4 square inches. Use the new formula to find the height of the triangle in inches and in centimeters.

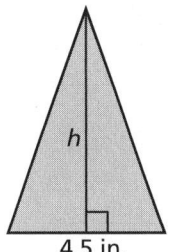

Find the slope and y-intercept of the graph of the linear equation. Then sketch its graph.

5. $y = 1.5x + 1$

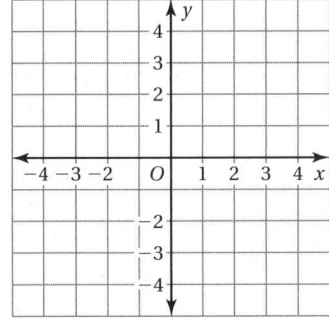

6. $3x + 5y = 1$

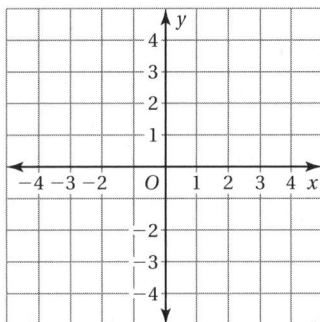

7. The equation $3.50x + 1.50y = 21$ represents the cost for a family to attend a play where x is the number of adults and y is the number of children. Find the intercepts and interpret the meaning of each one.

Solve the system.

8. $y = \dfrac{3}{2}x + 2$
   $y - \dfrac{1}{2}x = \dfrac{1}{2}$

9. $y - \dfrac{4}{3}x = 2.5$
   $3y = 4x - 2$

10. $y = x^2 + 5x$
    $y = 2x + 10$

11. It costs $0.07 to send a text message and $0.12 to send a picture on your cell phone. You spend $3.38 and send twice as many text messages as pictures. How many text messages did you send?

Answers

1. _____
2. _____
3. _____
   _____
4. a. _____
   _____
   b. _____
   _____
5. _____
   _____
   See left.
6. _____
   _____
   See left.
7. _____
   _____
   _____
8. _____
9. _____
10. _____
11. _____

Test 2 End-of-Course Test (continued)

Write an equation of the line in slope-intercept form.

12. the line passing through $(-1.5, 2)$ and $(1, -1)$

13. the line with slope -2 and passing through $(3, 1)$

Determine the domain and range of the function.

14.
x	−2.5	−1	0	1.2	2
y	−3	−1	1	3	5

15.
x	−1.5	−0.5	0	0.5	1.5
y	4	2	0	2	4

16. The input of one function is volume in liters. The input of another function is number of cars. Which function has a continuous domain? Explain.

17. The table shows the cost y (in dollars) of x peaches.

Peaches, x	0	4	8	12
Cost, y	0	3	6	9

a. Graph the data.

b. Is the domain discrete or continuous?

c. Write a linear function that relates y to x.

d. What is the cost of six peaches?

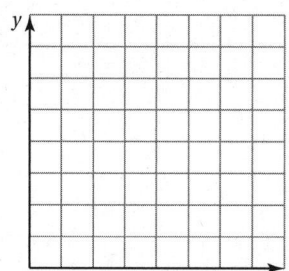

Answers

12. _____
13. _____
14. _____
    _____
15. _____
    _____
16. _____
    _____
17. a. _See left._
    b. _____
    c. _____
    d. _____
18. _____
19. _____
20. _____
21. _____
22. _____
23. _____
24. _____
25. _____
26. _____
27. _____

Simplify the expression.

18. $\left(x^4 y^{-3}\right)^{-2}$

19. $\dfrac{32a^2 b^{-1}}{12 a^5 b^3}$

20. $\sqrt[3]{343}$

21. $1000^{4/3}$

22. $(p^2 - p) - (7p^2 + 2p - 1)$

23. $(3n - 1)^2$

24. $\dfrac{c^2 - 9}{8c^3} \div \dfrac{c^2 - 6c + 9}{24c}$

25. $\dfrac{8}{u + 1} + \dfrac{u - 3}{u}$

Factor the polynomial.

26. $a^2 - 3a - 54$

27. $f^3 + 14f^2 + 49f$

Name_____ Date_____

Test 2 End-of-Course Test (continued)

28. A ladder is placed against the side of a house. The top of the ladder is 24 feet above the ground. The base of the ladder is 7 feet away from the house. Find the length of the ladder.

29. The data represent the weights (in pounds) of dogs at a dog show.

    7, 30, 37, 40, 40, 50, 45, 35, 32, 40, 55, 90

    a. Find the mean, median, mode, and range.

    b. Make a box-and-whisker plot for the data.

Choose an appropriate display for the situation. Explain your reasoning.

30. The percent of students who chose red, green, blue, yellow, or another color as their favorite color

31. The average cost of a movie ticket over the last twenty years

32. The table shows the number of years of college education and hourly earnings (in dollars) for several people.

Years, x	0	1	3	5	6
Hourly earnings, y	6	9	15	26	31

    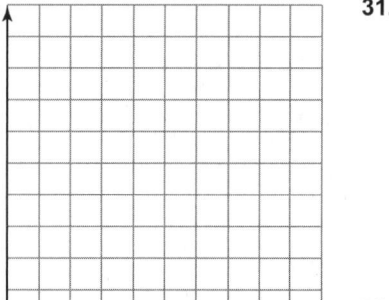

    a. Make a scatter plot of the data and draw a line of fit.

    b. Write an equation for the line of fit.

    c. Predict the hourly wage for a person with four years of college education.

Write the word sentence as an inequality.

33. 3.2 less than a number t is at most 7.5.

34. A number m multiplied by $\dfrac{4}{7}$ is greater than $\dfrac{12}{5}$.

Solve the inequality. Graph the solution.

35. $x + 2.5 > -4.3$

36. $\dfrac{1}{2}x + \dfrac{1}{3} \le \dfrac{2}{3}$

Answers

28. _____

29. a. _____

_____

_____

_____

b. See left.

30. _____

_____

_____

_____

_____

31. _____

_____

_____

_____

_____

_____

32. a. See left.

b. _____

c. _____

33. _____

34. _____

35. _____

 See left.

36. _____

 See left.

Test 2 End-of-Course Test (continued)

37. If you spend at least $50 (including shipping) at an online store, you receive a $10 gift card. You want to buy CDs that cost $12.50 each. If shipping costs $5, write and solve an inequality to find the number of CDs you must buy to receive the gift card.

Solve the equation. Check your solution(s).

38. $3^{x+4} = 27$

39. $w^2 - 25 = 0$

40. $x^2 + 6x - 27 = 0$

41. $d^2 + 3d + 7 = 0$

42. $\sqrt{z + 5} + 4 = 6$

43. $\dfrac{y}{y + 4} = \dfrac{5}{y - 3}$

Graph the function. Describe the domain and range.

44. $y = -(x + 5)^2 - 2$

45. $y = \dfrac{3}{4}x^2$

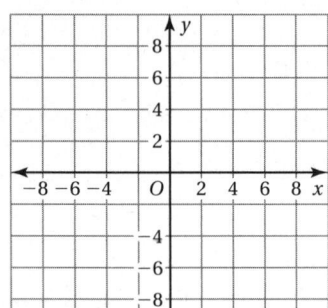
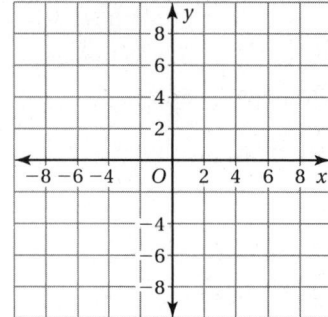

46. $y = \sqrt{x - 3} + 1$

47. $y = \dfrac{1}{x} + 3$

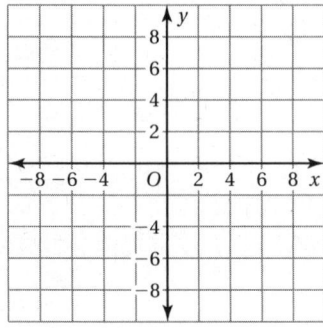
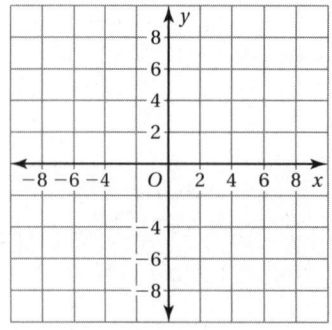

Answers

37. _____

38. _____
39. _____
40. _____
41. _____
42. _____
43. _____
44. _____See left._____

45. _____See left._____

46. _____See left._____

47. _____See left._____

Name_____ Date_____

Resource Gridded Response Answer Sheet

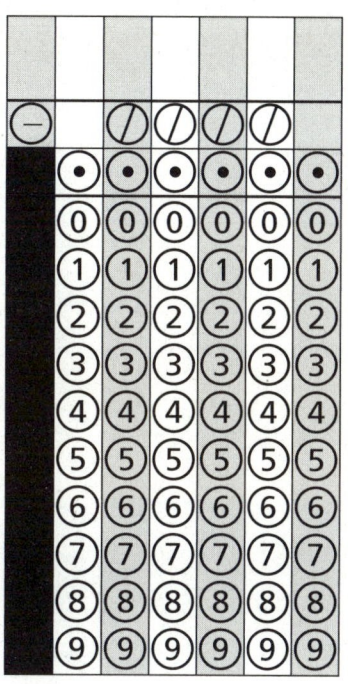
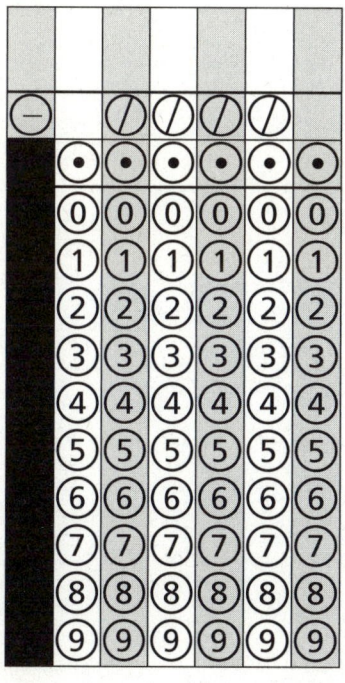
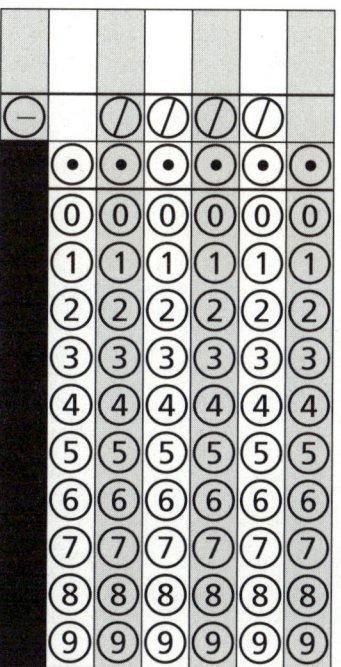

Name _____ Date _____

Resource Gridded Response Answer Sheet

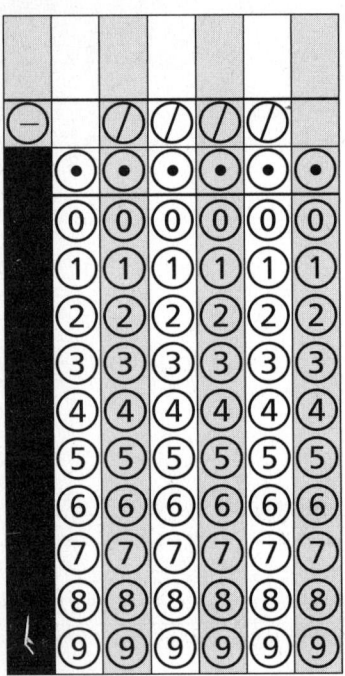
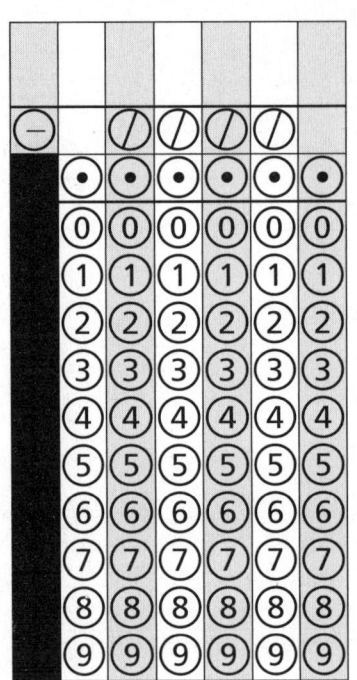
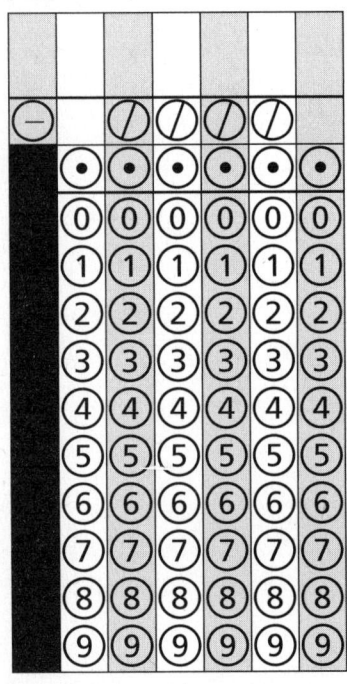
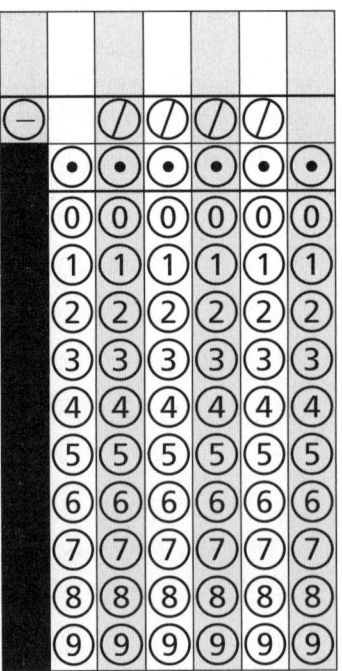

Answers

Pre-Course Test

1. no 2. yes 3. yes 4. $30.00
5. 1000 km 6. $(-2, 3)$ 7. $(-1, -4)$

8. 9.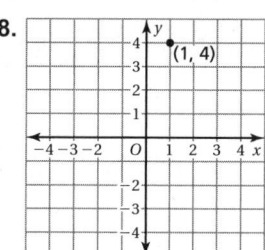

10. 7 11. 56 12. -15 13. 34
14. 41 15. 31 16. $\dfrac{37}{45}$
17. $-\dfrac{7}{6}$ or $-1\dfrac{1}{6}$ 18. $\dfrac{13}{50}$
19. $-\dfrac{17}{24}$ 20. -44.361 21. -2.54
22. $x = 7$ 23. $x = -8$ 24. $x = -7$
25. no solution 26. $x = -1$
27. infinitely many solutions
28. $6x + 3 - 3 = 27 - 3$ Subtraction Prop. of Equality
    $\dfrac{6x}{3} = \dfrac{24}{3}$ Division Prop. of Equality
    $2x = 8$

29. 11 m 30. $4\sqrt{3}$ in. 31. 0.75
32. 0.3125 33. 5.25
34. no; This sample favors people who like to wear red.

35.
Stem	Leaf
7	6 7 7 8 9
8	0 2 2 4 4 7
9	0 5 7
10	0

Key: 7 | 6 = 76

36. slope: 3; y-intercept: -8
37. 5 38. -2 39. $3\sqrt{6}$
40. 36 41. $81d^4$ 42. $\dfrac{1}{4}$

Chapter 1

1.1–1.2 Quiz

1. $c = 4\dfrac{1}{3}$ 2. $x = -2$ 3. $s = 1.2$
4. $r = 4$ 5. $d = 9$ 6. $m = 13\dfrac{1}{3}$
7. $n = 0$ 8. $q = 8$ 9. $x = 150$
10. $x = 160$ 11. $13
12. 15 ft, 30 ft, 45 ft, 60 ft
13. $d = 60t$; 6 hours 14. 11 cars

1.3–1.4 Quiz

1. $p = -4$ 2. $y = 0$
3. $k = 2$ or $k = -12$

4. $z = -5$ or $z = -11$

5. $y = \dfrac{5}{4}x - \dfrac{5}{2}$ 6. $y = 3x - 7$
7. $h = \dfrac{3V}{\pi r^2}$ 8. $b = \dfrac{2A}{h}$
9. at your school 10. $h = \dfrac{2A}{b + B}$; 16 ft
11. Route that passes the mall: 11 miles
    Route that passes the theater: 13 miles
12. $h = \dfrac{S - 2\pi r^2}{2\pi r}$

Test A

1. $y = 21$ 2. $r = -3.6$ 3. $x = 6$
4. $f = -20$ 5. $p = 7$ 6. $g = \dfrac{2}{3}$
7. $x = -3$ 8. $r = -\dfrac{8}{3}$ 9. $w = 18$

Answers

10. $q = -2$

11. $d = 4$ or $d = 1$

[number line with points at 1 and 4, range −4 to 4]

12. $t = -1$ or $t = -5$

[number line with points at −5 and −1, range −6 to 2]

13. $y = 3 - \dfrac{2}{5}x$

14. $y = \dfrac{4}{3} - \dfrac{1}{2}x$

15. $y = 0.5x - 2$

16. $y = 8x + 20$

17. a. $R = P + C$ **b.** $870

18. $h = \dfrac{V}{w\ell}$

19. $p = s + 0.2t$

20. $s = \dfrac{Z}{L}$

21. $T = \dfrac{PV}{nR}$

22. $160 = g + 24$; $g = 136$; 136 students

23. $5x = 12.50$; $x = 2.5$; 2.5 h

24. $395 = 24n + 35$; $24n = 360$; $n = 15$; 15 birthday cakes

25. $x = 25$

Test B

1. $d = 14$
2. $x = \dfrac{11}{15}$
3. $n = -\dfrac{5}{4}$

4. $s = -\pi$
5. $x = -2$
6. $w = -15$

7. $a = -1$
8. $g = \dfrac{2}{3}$
9. $x = 2$

10. $j = 4$

11. $k = 9$ or $k = 1$

[number line with points at 1 and 9]

12. no solution

13. $y = \dfrac{5}{3}x - \dfrac{2\pi}{3}$

14. $y = 1.6x - 2$

15. $y = 1.5x + 0.5$

16. $y = -\dfrac{1}{2}x + \dfrac{3}{2}$

17. a. $P = \dfrac{I}{rt}$ **b.** $500

18. $c = \dfrac{3i}{e}$

19. $w = \dfrac{P}{2} - \ell$

20. $R = \dfrac{V}{I}$

21. $h = \dfrac{S - 3\pi r^2}{2\pi r}$

22. $c + 11.50 = 47$; $c = 35.50$; $35.50

23. $3.50x = 31.50$; $x = 9$; 9 video games

24. $5x + 8.50 = 38.45$; $x = 5.99$; $5.99

25. a. $p = \dfrac{C}{1.06}$ **b.** $59

26. 43

Alternative Assessment

1. a. *Sample answer:*

Let the digits be t and w. Sum 1 is $t + w$. The two-digit numbers are "tw" and "wt" or $10t + w$ and $10w + t$. Sum 2 is $10t + w + 10w + t$, or $11t + 11w$.

$$Q = \dfrac{\text{Sum 2}}{\text{Sum 1}}$$
$$= \dfrac{11t + 11w}{t + w}$$
$$= \dfrac{11(t + w)}{t + w}$$
$$= 11$$

b. *Sample answer:* Let x be the number.

$$7x + 2 = 2x + 7$$
$$5x = 5$$
$$x = 1$$

Equation to show why it always works:

$$ax + b = bx + a$$
$$ax - bx = a - b$$
$$(a - b)x = a - b$$
$$x = 1$$

c. *Sample answer:*

$$\dfrac{4x + 8}{4} = n$$
$$x + 2 = n$$
$$x = n - 2$$

To find the original number choice, subtract 2 from the number given to you.

Answers

2. a. *Sample answer:* Jake ran farther, because $2\frac{1}{2}$ miles is 13,200 feet and that is twice the 6600 feet that Cal ran.

b. *Sample answer:* Both boys ran at the same rate: 5 miles per hour or 26,400 feet per hour.

Chapter 2

2.1–2.4 Quiz

1. 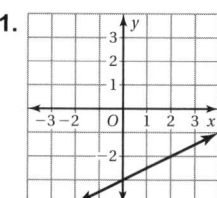 **2.**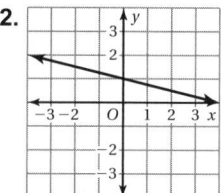

3. $\dfrac{4}{5}$ **4.** $-\dfrac{1}{3}$ **5.** $-\dfrac{5}{4}$

6. slope: -4; y-intercept: -6

7. slope: $\dfrac{1}{2}$; y-intercept: $-\dfrac{1}{3}$

8. x-intercept: 8; y-intercept: -6

9. x-intercept: -2; y-intercept: 4

10. $y = 1.75x + 2.5$

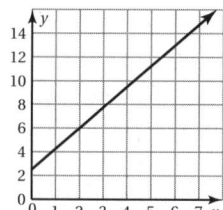

11. a.

[graph]

b. The x-intercept shows that you can buy 12 sandwiches if you don't buy any beverages. The y-intercept shows that you can buy 33 beverages if you don't buy any sandwiches.

2.5–2.7 Quiz

1. $y = 2x - 1$ **2.** $y = -\dfrac{1}{3}x + 1$

3. $y - 2 = -2(x - 1)$ **4.** $y + 2 = \dfrac{1}{4}(x - 4)$

5. $y + 2 = 1(x + 3)$ **6.** $y + 1 = -\dfrac{1}{3}(x - 9)$

7. $y = -1$ **8.** $y = 2x + 1$

9. a. $y = 6x - 27$ **b.** $y = -\dfrac{1}{6}x - \dfrac{7}{3}$

10. $y = 0.25x + 21$

11. a. $y = 125x + 120$

b. The y-intercept is the initial deposit.

12. $y = -2x + 10$

Test A

1. *Sample answer:*

x	0	4
$y = \dfrac{1}{2}x$	0	2

Sample answer: $(2, 1)$

2. *Sample answer:*

x	1	-1
$y = x + 3$	4	2

Sample answer: $(0, 3)$

Graph for Exercises 1 and 2

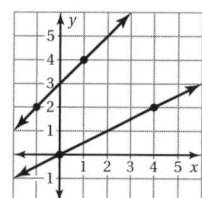

3. $y = -\dfrac{1}{4}x - 3$ **4.** $y = \dfrac{2}{3}x - 1$

5. 0 **6.** $\dfrac{1}{2}$

7. the sliding pole, because $\dfrac{5}{3} > \dfrac{3}{2}$

8. *Sample response:* $y = 2x + 1$

Answers

9. 3; −6   10. $-\dfrac{3}{4}$; −5   11. $\dfrac{7}{9}$; $-3\dfrac{1}{3}$

12. The y-intercept, −12, is the depth (12 m) at which the submarine starts at time 0. The slope −8 is the speed at which it descends, −8 m/min.

13. 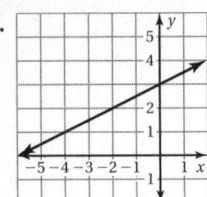   14.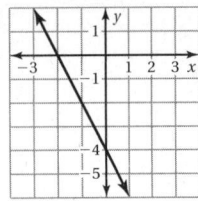

15. a. $6x + y = 9$

    b. 9; The distance from home at which you start at time 0.

    c. $1\dfrac{1}{2}$; The time after which you arrive home, in hours.

16. $y = \dfrac{3}{2}x + 1$ 17. $y = x + 4$

18. $y = -2x + 1$ 19. $y = -\dfrac{1}{2}x + 9$

20. a. slope = $\dfrac{4}{5}$; the kite rises 4 feet every 5 seconds (or the kite rises 0.8 ft per sec).

    b. $y = \dfrac{4}{5}x + 4$

    c. 16 ft

    d. When you first let out the string, the height of the kite is 4 feet.

Test B

1. $y = -\dfrac{3}{2}x - 2$ 2. $y = \dfrac{3}{4}x + 1$

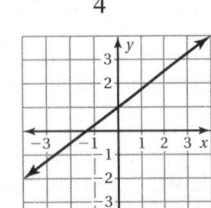

3. 2   4. $\dfrac{1}{3}$

5. The first hill is steeper; $\dfrac{2}{10} > \dfrac{2}{15}$.

6. The top and middle lines have the same slope, $\dfrac{2}{3}$, and are therefore parallel. The bottom line has a slope of $\dfrac{3}{4}$.

7. −2; −1   8. $\dfrac{1}{3}$; 0   9. $\dfrac{3}{4}$; −2

10. The x-intercept is the x-coordinate of the point where $y = 0$. Substitute 0 for y in the equation and solve. The x-intercept is $\dfrac{1}{2}$.

11. x-intercept, 2; y-intercept, −3

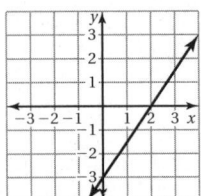

12. x-intercept, −1; y-intercept, 2

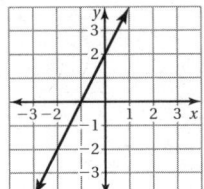

13. a. $15x + y = 90$

    b. 90, the amount owed at the start when $x = 0$.

    c. 6, the number of weeks needed to reduce the amount owed to 0.

14. $y = \dfrac{2}{3}x$ 15. $y = 3$

16. $y = \dfrac{5}{2}x + 7$ 17. $y = -3x + 22$

18. a. slope = −5; the temperature drops 5°F every 1000 feet.

    b. $y = -5x + 75$

    c. The x-intercept is 15; the temperature is 0°F at an altitude of 15,000 feet.

    d. 20°F

Answers

Alternative Assessment

1. a. $2.5t + 5m = 20$

b.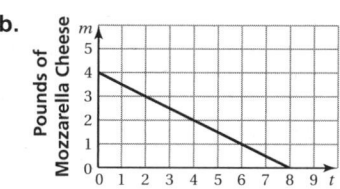

The m-intercept shows that Sophia can buy 4 pounds of mozzarella cheese if she doesn't buy any tomatoes. The t-intercept shows that Sophia can buy 8 pounds of tomatoes if she doesn't buy any mozzarella cheese.

c. She can buy 6 pounds of tomatoes and 1 pound of cheese, 4 pounds of tomatoes and 2 pounds of cheese, or 2 pounds of tomatoes and 3 pounds of cheese.

d. She bought 4 pounds of tomatoes and 2 pounds of mozzarella cheese.

2. a. 1.5

b. The tree grows at a rate of 1.5 feet per year.

c. $h = 1.5t + 3$

d. 48 feet

e. 38 years

f. The graph would have a steeper slope.

Chapter 3

3.1–3.3 Quiz

1. $c - 12 > 4$ **2.** $y + 3.6 \leq 9.5$

3. yes **4.** no

5. 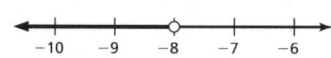 **6.**

7.

8. $y \leq 3$

9. $z \geq 3$

10. $d < -8$

11. $-5x \geq 35;\ x \leq -7$ **12.** $\dfrac{x}{3} \leq 12;\ x \leq 36$

13. $x \geq 65$ **14.** $b = 7$

15. $x + 1.5 \leq 14.5;\ x \leq 13$ gallons

16. $300x \geq 1500;\ x \geq 5$ levels

3.4–3.5 Quiz

1. $t < 3$ **2.** $w \geq 2$
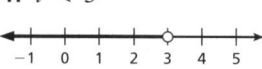

3. $2 < j \leq 5$ **4.** $n \geq 4$ or $n \leq -2$
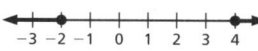

5. $c < 9$ or $c \geq 12$ **6.** $k > -1$ or $k < -2$
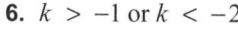
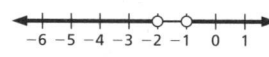

7. 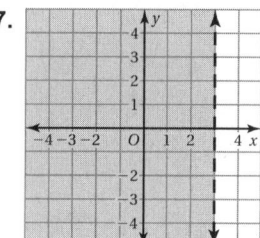 **8.**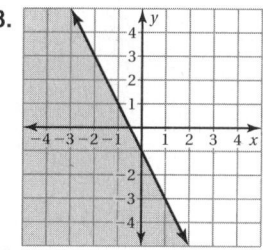

9. $350 - 50x \geq 100;\ x \leq 5$

10. $0.5x + 0.25y \leq 5$, where x is the number of dart games played and y is the number of ring toss games played.

Test A

1. $x > -1$ **2.** $x \leq -2$

3. $n \geq -3$ **4.** $q + 7 < 45$

5. $x \div (-1) \geq -4$ **6.** $a > 10$

7. $c \geq 3$ **8.** no **9.** yes

10. Cools when $T > -10$; off when $T < -16$; *Sample answer:* cools when $T = -8$, off when $T = -18$.

11. a. $x + 95 \leq 800,\ x \leq 705$; up to 705 lb

b. $y + 280 \leq 705,\ y \leq 425$; up to 425 lb

c. No; $470 > 425$ and is outside the safe limit for weight that can be added.

Answers

12. $x > 10$ **13.** $m \leq -6$ **14.** $y > 1\frac{1}{3}$

15. $p < -15$ **16.** $z \geq \frac{1}{2}$ **17.** $1 < t < 4$

18. $x \leq 5$

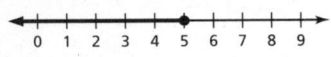

19. $y < -10$

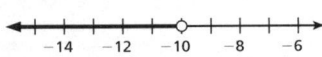

20. $x \geq 0$ or $x \leq -1$

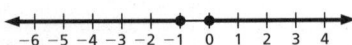

21. 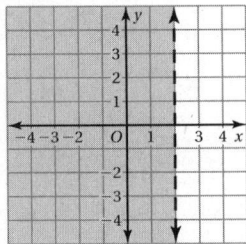 **22.**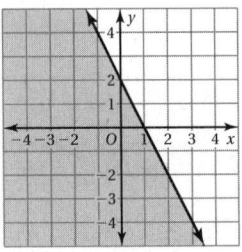

23. $t \leq 70$, where t is time in minutes

Test B

1. $x < \frac{1}{4}$ 2. $n \leq 8$

3. $m - 3 > -4$ 4. $16j \geq -2$

5. $2q - 1 < 5$ 6. $a \div 2 \leq 6$

7. $s \geq 60$ 8. $c \leq 35$

9. yes 10. no

11. **a.** $t + 55 \leq 100$, $t \leq 45$; up to 45 sec
    b. $t + 55 \leq 70$, $t \leq 15$; up to 15 sec
    c. no; yes; 32 is less than 45 but not less than 15.

12. $2x + 5 \leq 30$, $x \leq 12.5$

13. $b > -1$ 14. $x \leq -1.5$

15. $c \leq 5$ 16. $p > -15$

17. $w \geq -3$ 18. $n < 0$ and $n > -4$

19. $w \geq 4$

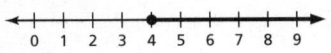

20. $m > 40$

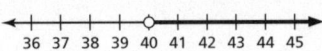

21. $z \geq 1\frac{2}{3}$

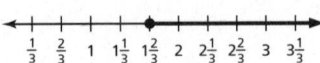

22. $x < -2$ or $x > -1$

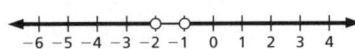

23. 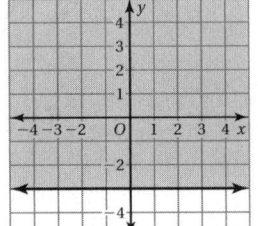 **24.**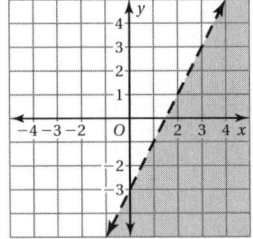

25. 12 or less; the amount is a maximum. $12x \leq 150$ leads to $x \leq 12.5$, and you cannot buy half a book, so $x \leq 12$, or x is at most 12; this is a maximum amount.

Alternative Assessment

1. **a.** $5 + z \leq 8$, $5 - q \geq 8$
   b. $z \leq 3$, $q \leq -3$
   c. $z \leq 3$

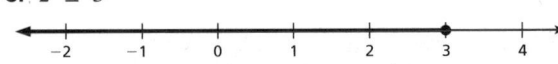

   $q \leq -3$

   Sample answer: The circles are closed because 3 and −3 are part of the solutions.

   d. *Sample answer:* Both solutions have the $\leq$ inequality symbol, both solutions have an absolute value of 3, but one is positive and the other is negative.

   e. *Sample answer:* Every morning Jessie, Kate, and Lois try to meet to walk a mile. Sometimes they are not all able to meet. How many miles altogether does the group walk each day?

2. **a.** $2t > 4$, $-2v < 4$
   b. $t > 2$, $v > -2$

Answers

c. $t > 2$

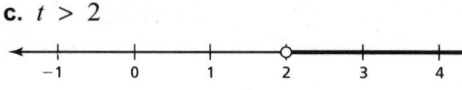

$v > -2$

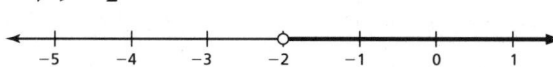

Sample answer: The circles are open because 2 and -2 are not part of the solutions.

d. *Sample answer:* In the graphs, the direction of the solution is the same, but one graph starts at -2 and the other at 2.

3. a. $3m - 1 \geq 2, w + 2 \leq 3$

b. $m \geq 1, w \leq 1$

c. $m \geq 1$

$w \leq 1$

Sample answer: The circles are closed because 1 is a part of both solutions.

d. *Sample answer:* Both solutions include the value 1, so both graphs have closed circles at 1. The direction of the solution is different for the two graphs, so one is like a mirror image of the other.

Chapter 4

4.1–4.2 Quiz

1. A; $(2, 3)$ **2.** B; $(2, -3)$

3. $(-4, 2)$ **4.** $(-4, -1)$

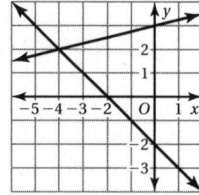 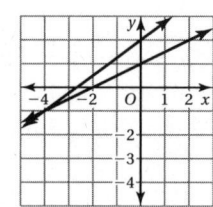

5. $(-2, -1)$ **6.** $(1, 2)$

7. $(12, 6)$ **8.** 15 boys; 25 girls

9. length = 13 feet; width = 5 feet

4.3–4.5 Quiz

1. $(-4, 2)$ **2.** $(1, -7)$ **3.** $(-6, 3)$

4. infinitely many solutions

5. no solutions **6.** $(-3, 0)$

7. $x = 1$ **8.** $x = 2$

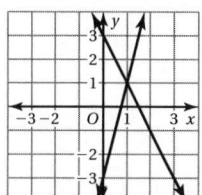 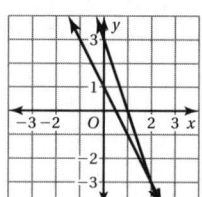

9.

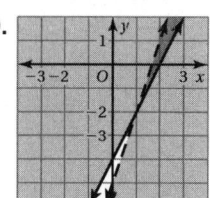

10. $y > -\dfrac{1}{3}x - 2$

$y \leq x + 2$

11. 20 scented candles; 8 unscented candles

12. $x + y \geq 4$

$x > \dfrac{1}{2}y + 1$

Test A

1. B; $(1, 1)$ **2.** A; $(-1, -1)$

3. $(1, -1)$ **4.** $(1, 3)$

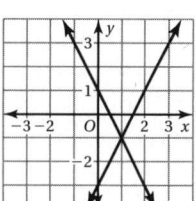 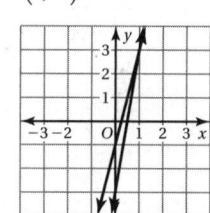

5. $(3, 5)$ **6.** $(1, 7)$ **7.** $(-3, 1)$

8. 19 red marbles, 8 blue marbles

9. 36 apples, 24 oranges

10. $(1, -3)$ **11.** $(-20, 2)$ **12.** $(30, 10)$

13. infinitely many solutions; The two equations are equivalent.

14. no solution; The two equations represent lines with the same slope but different y-intercepts.

15. one solution; The two equations represent lines with different slopes. Because the two lines have different slopes, they are not parallel and must intersect at a point.

Answers

16. $x = -1$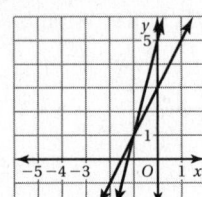

17. $x = -1$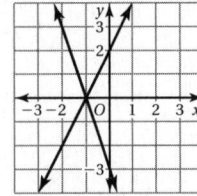

18.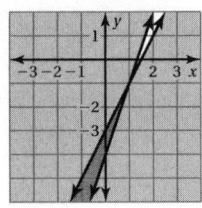

19.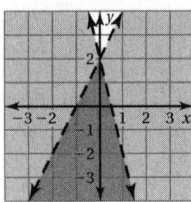

20. $6

Test B

1. $(-3, 0)$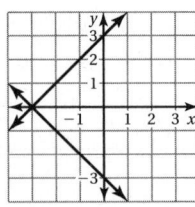

2. $(3, 3)$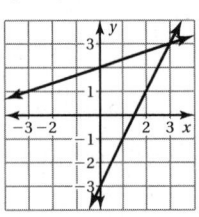

3. $(1, 1)$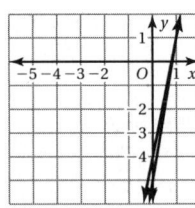

4. $(3, 1)$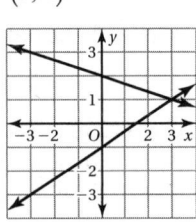

5. $(3, 6)$ **6.** $(1, 6)$ **7.** $(-3, 3)$

8. 20 red pens; 4 blue pens

9. 15 orange fish; 16 red fish

10. $110°$; $70°$

11. $(-1, -5)$ **12.** $(4, 10)$ **13.** $(-1, -8)$

14. one solution; The two equations represent lines with different slopes. Because the two lines have different slopes, they are not parallel and must intersect at a point.

15. infinitely many solutions; The two equations are equivalent.

16. no solution; The two equations represent lines with the same slope but different *y*-intercepts.

17. $x = 2$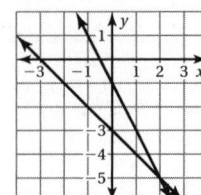

18. $x = -1$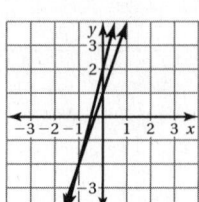

19.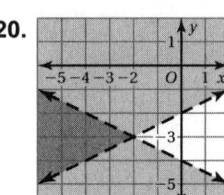

20.

21. a. $0.75x + 1.25y \leq 3.75$

$x + y \geq 3$

b. yes; Both inequalities are true when you substitute $x = 3$ and $y = 1$.

Alternative Assessment

1. a.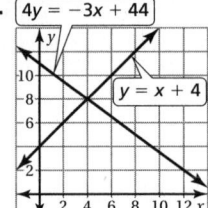

The solution to the system of linear equations is $(4, 8)$.

b. *Sample answer:* I drew a line that passed through both the origin and $(4, 8)$.

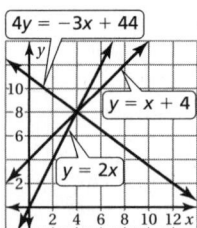

c. $y = 2x$

Answers

d. *Sample answer:* $y = 6x + 16$; $y = -2x + 16$

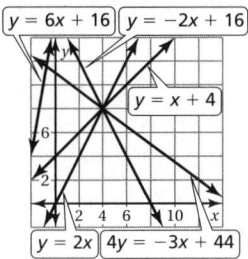

e. *Sample answer:*

$y = x + 4$
$8 \stackrel{?}{=} 4 + 4$
$8 = 8$

$4y = -3x + 44$
$4(8) \stackrel{?}{=} -3(4) + 44$
$32 \stackrel{?}{=} -12 + 44$
$32 = 32$

$y = 2x$
$8 \stackrel{?}{=} (2)4$
$8 = 8$

$y = 6x - 16$
$8 \stackrel{?}{=} 6(4) - 16$
$8 \stackrel{?}{=} 24 - 16$
$8 = 8$

$y = -2x + 16$
$8 \stackrel{?}{=} -2(4) + 16$
$8 \stackrel{?}{=} 8 + 16$
$8 = 8$

f. *Sample answer:* equations used: $y = 2x$ and $y = x + 4$. Yolanda says she has twice as many T-shirts as her younger brother, Xavier. Xavier said that she only has 4 more than he does. How many T-shirts do Yolanda and Xavier have?

2. a. *Sample answer:* Both equations are linear and both are in the same variables, so they form a system of linear equations.

b.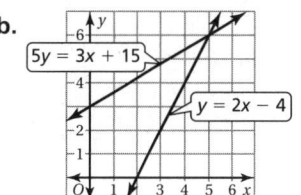

The solution is $(5, 6)$.

$y = 2x - 4$ and $5y = 3x + 15$, so

$2x - 4 = \dfrac{3}{5}x + 3$

$\dfrac{7}{5}x = 7$

$x = 5$

$y = 2x - 4$
$ = 2(5) - 4$
$ = 10 - 4$
$ = 6$

The solution is $(5, 6)$.

c. *Sample answer:* The slope of $y = 2x - 4$ is 2. The slope of $5y = 3x + 15$ is $\dfrac{3}{5}$. To find slope on the graph, find the ratio of rise over run by counting the number of units y changes between two points, and the number of units x changes. To find slope in an equation, when the equation is solved for y, the coefficient of x is the slope.

d. The x-intercept of $y = 2x - 4$ is 2 and the y-intercept is -4. The x-intercept of $5y = 3x + 15$ is -5 and the y-intercept is 3. To find these values graphically, examine the graphs to see where the line for the equation crosses the y-axis to find the y-intercept and where it crosses the x-axis to find the x-intercept. To find these values algebraically, find the value of y when x is zero to find the y-intercept and find the value of x when y is zero to find the x-intercept.

e. $2x - y = 4$; $3x - 5y = -15$

Answers

Chapter 5

5.1–5.3 Quiz

1. domain: $-1, 0, 1, 2$; range: $-1, 0, 1, 2$

2. domain: $-2, -1, 0, 1, 2$; range: $-2, -1, 2$

3. discrete

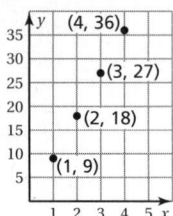

4. $y = 2x - 3$ 5. $y = -\dfrac{1}{2}x + 4$

6. discrete

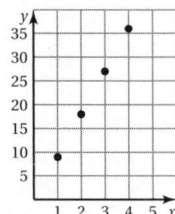

7. a. $g = -\dfrac{1}{20}m + 20$ b. 14 gallons

5.4–5.6 Quiz

1. $f(-5) = -9$; $f(0) = -4$; $f(10) = 6$

2. $g(-5) = 20$; $g(0) = 5$; $g(10) = -25$

3. $h(-5) = 1$; $h(0) = 2$; $h(10) = 4$

4. The graph of g is the translation 3 units up of the graph of f.

5. The graph of h is the translation 2 units down of the graph of f.

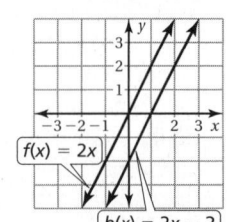

6. linear   7. nonlinear

8. $a_n = n + 10$; $a_{20} = 30$

9. $a_n = -3n$; $a_{20} = -60$

10. $20

11. nonlinear; As t increases by 1, v increases by different amounts.

12. $a_n = 0.50n + 0.40$

Test A

1. 1, 2, 4   2. $y = -2x + 6$

3. 0, 2, 4, 6

4.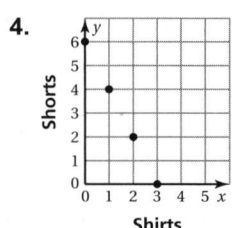

5. discrete

6.
x	-1	0	1	2
y	-4	1	6	11

7. discrete   8. continuous   9. $y = 2x + 2$

10. $y = 4x$ 11. continuous

12. $f(-4) = 7$; $f(0) = 11$; $f(8) = 19$

13. $g(-4) = 10$; $g(0) = 6$; $g(8) = -2$

14. The graph of g is the translation 5 units down of the graph of f.

15. a. $120   b. 25 hours

16. nonlinear; The graph is not a line.

17. linear; The graph is a line.

18. 24

19. $a_n = 3n + 5$; $a_{25} = 80$

20. $a_n = -5n + 4$; $a_{25} = -121$

Test B

1. domain: $-2, -1, 1, 3$; range: 1, 2, 3

2. 0, 4, 8   3. discrete

4. $y = -0.5x + 4$

Answers

5.

x	−1	0	1	2
y	5.2	4	2.8	1.6

6. Yes, you can have 15 skateboards.

7. $P = 4s$; linear

8. $y = x + 1$

9. $y = -\dfrac{x}{2}$

10. a.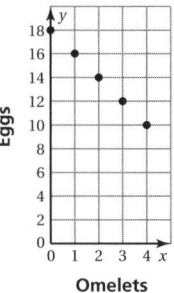

b. discrete

11. $f(-5) = -16$; $f(0) = -6$; $f(10) = 14$

12. $g(-5) = -6$; $g(0) = -7$; $g(10) = -9$

13. The graph of g is the translation $\dfrac{1}{2}$ unit up of the graph of f.

14. a. $f(x) = 15x$ b. 5 boxes

15. for Car A

16. nonlinear; The equation cannot be written in slope-intercept form.

17. linear; The equation can be written in slope-intercept form.

18. any positive value other than 52

19. $a_n = 10n - 9$; $a_{30} = 291$

20. $a_n = 5n - 20$; $a_{30} = 130$

Alternative Assessment

1. a.

$y = x + 2$	
x	y
2	4
4	6
6	8
8	10

$y = x - 2$	
x	y
2	0
4	2
6	4
8	6

$y = \dfrac{2}{x}$	
x	y
2	1
4	$\dfrac{1}{2}$
6	$\dfrac{1}{3}$
8	$\dfrac{1}{4}$

$y = \dfrac{x}{2}$	
x	y
2	1
4	2
6	3
8	4

$y = 2x$	
x	y
2	4
4	8
6	12
8	16

b.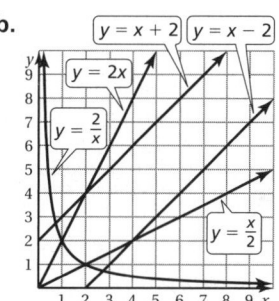

c. All the functions except $y = \dfrac{2}{x}$ are linear. The graph of $y = \dfrac{2}{x}$ is not a straight line. All of the others are straight lines, so they are all linear functions.

Answers

  d. $y = x + 2$: The domain is 2, 4, 6, and 8.
     The range is 4, 6, 8, and 10.
     $y = x - 2$: The domain is 2, 4, 6, and 8.
     The range is 0, 2, 4, and 6.
     $y = \dfrac{2}{x}$: The domain is 2, 4, 6, and 8.
     The range is $1, \dfrac{1}{2}, \dfrac{1}{3}$, and $\dfrac{1}{4}$.
     $y = \dfrac{x}{2}$: The domain is 2, 4, 6, and 8.
     The range is 1, 2, 3, and 4.
     $y = 2x$: The domain is 2, 4, 6, and 8.
     The range is 4, 8, 12, and 16.

2. **a.** Table 1 represents line W, Table 2 represents line F, and Table 3 represents line N.
   b. For Table 1 and line W: $y = 2x + 1$;
   for Table 2 and line F: $y = 3x$; and
   for Table 3 and line N: $y = x + 2$.
   c. *Sample answer:* For Table 3: Barbara and Mike are in the reading club. Every month last year, Barbara read 2 books more than Mike did. If Barbara read 5 books in April, how many books did Mike read in April?
   d. *Sample answer:* For Table 2: At the farmer's market, blueberries were selling for $3 a pound. How much did Roger pay for $1\dfrac{1}{2}$ pounds of blueberries?

Chapter 6

6.1–6.3 Quiz

1. $4\sqrt{6}$ 2. $\dfrac{-\sqrt{17}}{4}$ 3. $-5 - \sqrt{2}$

4. $\dfrac{1 + \sqrt{3}}{3}$ 5. $4\sqrt{3}$ 6. $3\sqrt{11}$

7. 4^5 8. $\dfrac{1}{36d^2}$ 9. $\dfrac{1}{n^3}$

10. $\dfrac{w^6}{27}$ 11. 5   12. 4

13. 243   14. 8   15. $42\sqrt{15}$ in.²

16. *Sample answers:*
    $a = \sqrt{8}, b = \sqrt{2}; a = 2\pi, b = \pi$

17. $7x^4y^2$ mi/h   18. $r = 3$ ft

6.4–6.7 Quiz

1. exponential; Each successive term increases by the same factor.
2. linear; Each successive term increases by the same amount.

3.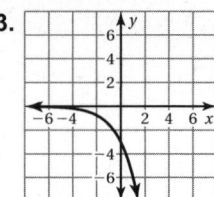
   Domain: all real numbers
   Range: $y < 0$

4.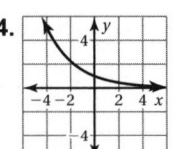
   Domain: all real numbers
   Range: $y > 0$

5. $x = -3$ 6. $x = 2$

7. exponential decay function
8. exponential growth function
9. 1250, 6250, 31,250   10. 56, −28, 14

11. $a_1 = 162, a_n = \dfrac{1}{3}a_{n-1}$

12. $a_1 = -15, a_n = a_{n-1} + 8$

13. **a.** $y = 35{,}000(0.86)^t$ **b.** $12,177.47

Test A

1. $2\sqrt{15}$ 2. $\dfrac{3\sqrt{3}}{4}$ 3. $3 - \sqrt{7}$

4. $\dfrac{3 + \sqrt{3}}{2}$ 5. n^{11} 6. $125x^3$

7. $\dfrac{1}{d^7}$ 8. $\dfrac{36}{x^{10}}$ 9. 5   10. 7

11. 81   12. 256   13. 36   14. 27

Answers

15.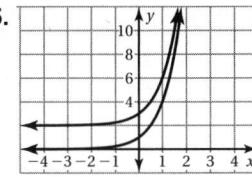

The domain is all real numbers. The range is all real numbers greater than 2. The graph is a translation of 2 units up.

16. $y = 7^x$ **17.** $y = -4(3)^x$

18. $x = -1$ **19.** $x = -2$

20. $y = 840(1.05)^x$

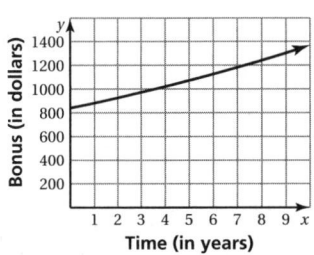

21. $y = 3000(1.025)^x$

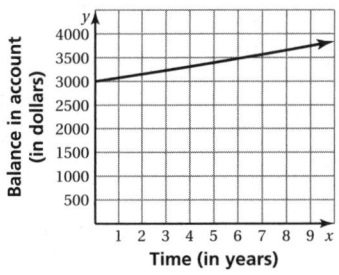

22. a. exponential decay
  b. $y = 18,000(0.85)^t$
  c. $4904.83

23. $a_n = 2(-3)^{n-1}$; 1458

24. $a_n = 50,500(0.1)^{n-1}$; 0.0505

25. Each side length is 7 yards.

Test B

1. $-6\sqrt{2}$ **2.** $\dfrac{5\sqrt{7}}{6}$ **3.** $\dfrac{5 + \sqrt{11}}{2}$

4. $\dfrac{-3 + \sqrt{3}}{2}$ **5.** $\dfrac{1}{a^5}$ **6.** $\dfrac{1}{64m^3}$

7. $\dfrac{9f^2}{4}$ **8.** $\dfrac{1}{c^9}$ **9.** 4 **10.** 11

11. 8 **12.** 27 **13.** -48 **14.** 8

15.

The domain is all real numbers. The range is all real numbers greater than 0. The graph is a translation of 2 units to the right.

16. $y = 2(6)^x$ **17.** $y = -5(4)^x$

18. $x = 3$ **19.** $x = -6$

20. $y = 15,000(1.085)^x$

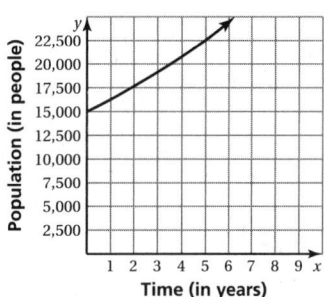

21. $y = 350(1.0475)^x$

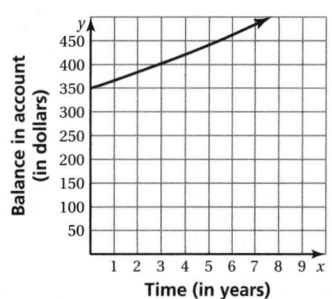

22. a. exponential decay **b.** 14% **c.** $5535.53

23. $a_n = -3(2)^{n-1}$; -192

24. $a_n = 4\left(\dfrac{1}{2}\right)^{n-1}$; $\dfrac{1}{16}$ **25.** $\dfrac{5}{6}$ yd

Answers

Alternative Assessment

1. a. $3\sqrt{10}$; The radicand, 10, has no perfect square other than 1.

   b. $-\dfrac{6\sqrt{2}}{17}$; The denominator does not have a radical, and the radicand, 2, has no perfect square other than 1.

   c. $2 - 2\sqrt{6}$; The radicand, 6, has no perfect square other than 1.

   d. x^5; The expression contains only positive exponents.

   e. $\dfrac{1}{y^4}$; The expression contains only positive exponents.

   f. $\dfrac{1}{d^6}$; The expression contains only positive exponents.

   g. $3\sqrt[3]{25}$; The radicand, 25, has no perfect cubes other than 1.

   h. 625; The e3xpression is an integer.

2. a. exponential; The *y*-values change by a factor of 1.5 as the *x*-values increase by 1.

   b. $y = 16\left(\dfrac{3}{2}\right)^t$

   c. 410 flowers

   d.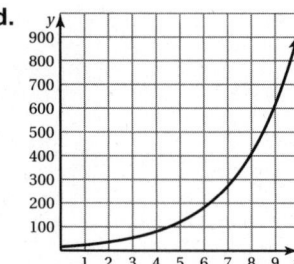

   e. domain: all real numbers greater than or equal to zero; range: all real numbers greater than or equal to 16.

   f. $(0, 16)$; The number of flowers initially planted was 16.

Chapter 7

7.1–7.4 Quiz

1. 0 2. 2 3. 6

4. $4y^6 + 5y^4$; 6; binomial

5. $r^3 + 2r - 9$; 3; trinomial

6. $7k^4 + 3k - 6$ 7. $7p^2 - 3p + 5$

8. $3m^2 + 4m - 13$

9. a. $12x + 110$ b. $290

10. $a^2 - 6a + 5$ 11. $6g^2 + 13g - 28$

12. $b^2 + 7b - 18$ 13. $20w^2 + 13m + 2$

14. $m^2 - 49$ 15. $9v^2 + 48v + 64$

16. a. $x^2 + 24x + 144$ b. 400 ft^2; 256 ft^2

7.5–7.9 Quiz

1. $3(p - 6)$ 2. $11(3b^2 + 2)$

3. $(k + 4)(k + 1)$ 4. $(p - 9)(p + 6)$

5. $2(2m + 1)(m + 3)$ 6. $(7a - 2)(a + 2)$

7. $(h + 10)^2$ 8. $(z - 6)^2$

9. $t = 0, t = -6$ 10. $g = -6, g = -3$

11. $a = -5, a = 2$ 12. $c = -4, c = 3$

13. $x = -6, x = 7$ 14. $c = -9, c = 4$

15. $w = -12, w = 12$ 16. $v = -5, v = 0$

17. 8 ft 18. length: 8 in., width: 4 in.

19. 3.75 sec

Test A

1. $3z^2 + 8z$; 2; binomial

2. $-w^4 - 3w + 6$; 4; trinomial

3. $7m^5$; 5; monomial

4. $9q + 7$ 5. $u^2 - 4u + 2$

6. $4a^2 - 6b^2 - 5ab$ 7. $g^2 + 13g + 42$

8. $b^2 + 7b - 18$ 9. $10t^2 - 57t + 54$

10. $12w^2 - 8w - 32$ 11. $m^2 - 49$

Answers

12. $9v^2 + 48v + 64$ **13.** $w^2 + 12w + 35$

14. $2a^2 + 9a - 5$

15. a. $28x + 48$ **b.** 160 in.^{2}

16. $5c(c^3 + 12)$ **17.** $(y - 3)(y - 7)$

18. $(a + 5)(a - 3)$ **19.** $(m - 7)(m + 4)$

20. $(w - 9)(w + 9)$ **21.** $(3x + 1)(2x - 3)$

22. $w = -3, w = 10$ **23.** $n = -4, n = 0$

24. $q = 0, q = 4$ **25.** $v = 2, v = 11$

26. $t = -7, t = 3$ **27.** $x = -4, x = \dfrac{1}{2}$

28. $u = -\dfrac{6}{5}, u = \dfrac{6}{5}$ **29.** $z = 12$

30. a. 1.25 sec   **b.** 6.25 ft

31. year 1 and year 4

32. a. $k = 8$ **b.** 30 ft^2; 80 ft^2

Test B

1. $-7h^3 + 12$; 3; binomial

2. $-6y^8 + 3y^2 - 4$; 8; trinomial

3. $1.6d^2$; 2; monomial

4. $5t^2 - 8t + 8$ **5.** $-11w^3 + 10w$

6. $2r^2 - 2rs - 6s^2$ **7.** $n^2 + 6n - 27$

8. $p^2 + 2p - 48$ **9.** $5b^2 - 38b + 21$

10. $16s^2 - 64$ **11.** $k^2 - 12k + 36$

12. $25y^2 + 110y + 121$

13. a. $10x^2 - 19x + 7$ **b.** 9 in.^{2}

14. a. $75r^2 + 150r + 75$ **b.** $\$90.75$ **c.** $\$29.25$

15. $4w(w^2 - 2)$ **16.** $(m - 13)(m - 3)$

17. $(a + 7)(a + 2)$ **18.** $(h + 16)(h - 2)$

19. $(8 + 3a)(8 - 3a)$ **20.** $3f(f + 3)(f - 4)$

21. $x = -3, x = 0$ **22.** $k = 0, k = 16$

23. $c = 3, c = 9$ **24.** $f = 4, f = 18$

25. $d = -13, d = 2$ **26.** $w = -\dfrac{1}{2}, w = 5$

27. $v = -\dfrac{2}{5}, v = \dfrac{2}{5}$ **28.** $s = 11$

29. a. 5 sec   **b.** 4 sec   **c.** 144 ft

30. year 2 and year 4

31. a. $y = 7$ **b.** 72 yd^2

Alternative Assessment

1. a. $2x^3 - 9x^2 - 5x$; degree: 3; trinomial

   b. $2x^3 - 6x^2 - 5x + 5$

   c. $2x^3 - 10x^2 - 9x + 1$

   d. $2x^4 - 7x^3 - 14x^2 - 5x$

   e. $2x^2 - 3x - 14 - \dfrac{42}{x - 3}$

   f. $x(x - 5)(2x + 1)$

   g. $x = -\dfrac{1}{2}, x = 0, x = 5$

2. a. Because $(a + b)^2 = (a + b)(a + b)$
   $= a^2 + 2ab + b^2$ and $a^2 + 2ab + b^2$
   $\neq a^2 + b^2$, $(a + b)^2 \neq a^2 + b^2$.

   Sample answer: When $a = 2$ and $b = 3$:

$(a + b)^2$	$a^2 + b^2$
$= (2 + 3)^2$	$= (2)^2 + (3)^2$
$= (5)^2$	$= 4 + 9$
$= 25$	$= 13$

   Because $25 \neq 13$, $(a + b)^2 \neq a^2 + b^2$

3. a. $2x - 3$ **b.** $8x - 12$ **c.** 68 ft

Answers

Chapter 8

8.1–8.3 Quiz

1. vertex: $(-3, 1)$; axis of symmetry: $x = -3$; domain is all real numbers; range is all real numbers greater than or equal to 1; when $x < -3$, y increases as x decreases; when $x > -3$, y increases as x increases

2. vertex: $(2, 3)$; axis of symmetry: $x = 2$; domain is all real numbers; range is all real numbers less than or equal to 3; when $x < 2$, y decreases as x decreases; when $x > 2$, y decreases as x increases

3.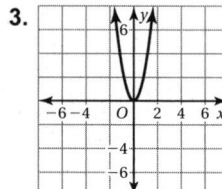

   Both graphs open up and have the same vertex, $(0, 0)$, and the same axis of symmetry, $x = 0$. The graph of $y = 3x^2$ is narrower than the graph of $y = x^2$.

4.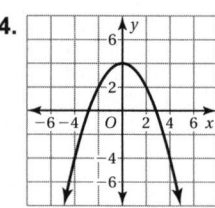

   Both graphs have the axis of symmetry, $x = 0$, but the graph of $y = -\frac{1}{2}x^2 + 4$ opens down. The graph of $y = -\frac{1}{2}x^2 + 4$ is wider than, a reflection in the x-axis, and a translation 4 units up of the graph of $y = x^2$.

5. 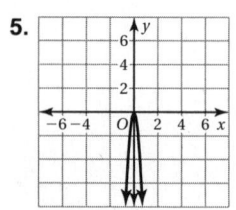   6.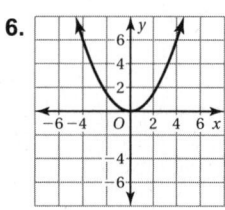

   focus: $\left(0, -\dfrac{1}{16}\right)$ focus: $\left(0, \dfrac{2}{3}\right)$

7. $y = \dfrac{1}{16}x^2$ 8. $y = \dfrac{1}{64}x^2$

8.4–8.5 Quiz

1. a. $x = -2$ b. $(-2, -7)$
2. a. $x = 4$ b. $(4, 1)$

3.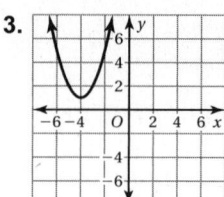

   The domain is all real numbers. The range is all real numbers greater than or equal to 1.

4.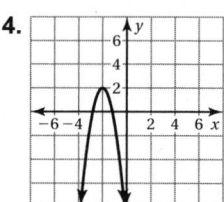

   The domain is all real numbers. The range is all real numbers less than or equal to 2.

5. minimum; $(-5, -57)$ 6. maximum; $(6, 4)$

7.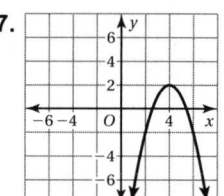

   The graph of $y = -(x - 4)^2 + 2$ is a reflection in the x-axis and a translation 4 units to the right and 2 units up of the graph of $y = x^2$.

8.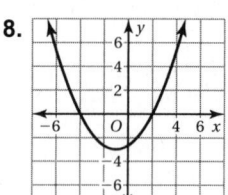

   The graph of $y = \dfrac{1}{3}(x + 1)^2 - 3$ is wider than, and is a translation 1 unit to the left and 3 units down of the graph $y = x^2$.

9. quadratic; $y = -\dfrac{1}{2}x^2$

10. The water reaches its maximum height of 58 feet, one second after it leaves the fountain.

Answers

Test A

1.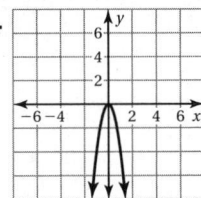

Both graphs have the same vertex, $(0, 0)$, and the same axis of symmetry, $x = 0$, but the graph of $y = -4x^2$ opens down. The graph of $y = -4x^2$ is narrower than and is a reflection in the x-axis of the graph of $y = x^2$.

2.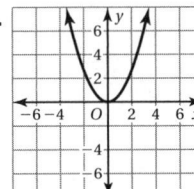

Both graphs open up and have the same vertex, $(0, 0)$, and the same axis of symmetry, $x = 0$. The graph of $y = \frac{2}{3}x^2$ is wider than the graph of $y = x^2$.

3.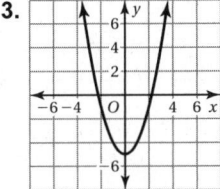

Both graphs have the same axis of symmetry, $x = 0$. The graph of $y = x^2 - 5$ is a translation 5 units down of the graph of $y = x^2$.

4.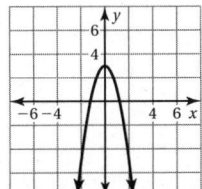

Both graphs have the axis of symmetry, $x = 0$, but the graph of $y = -2x^2 + 3$ opens down. The graph of $y = -2x^2 + 3$ is narrower than, a reflection in the x-axis, and a translation 3 units up of the graph of $y = x^2$.

5.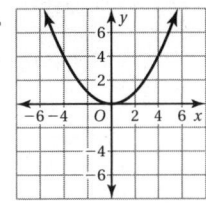

focus: $(0, 1)$

6.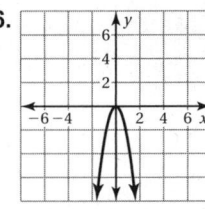

focus: $\left(0, -\dfrac{1}{12}\right)$

7.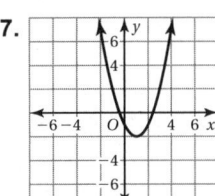

The domain is all real numbers. The range is all real number greater than or equal to -2.

8.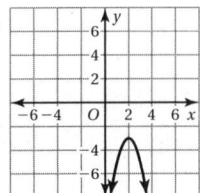

The domain is all real numbers. The range is all real number less than or equal to -3.

9.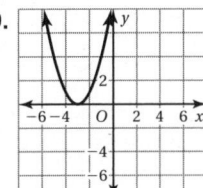

The graph of $y = (x + 3)^2$ is a translation 3 units to the left of the graph of $y = x^2$.

10.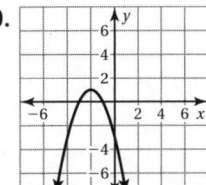

The graph of $y = -(x + 2)^2 + 1$ is a reflection in the x-axis and a translation 2 units to the right and 1 unit up of the graph of $y = x^2$.

11. $y = \dfrac{1}{8}x^2$

Answers

12. a.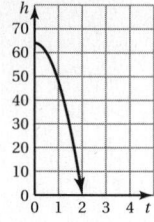

b. 2 seconds

13. quadratic; $y = \dfrac{1}{2}x^2$

14. a. quadratic **b.** $d = 0.75w^2$ **c.** 36.75 mi

Test B

1.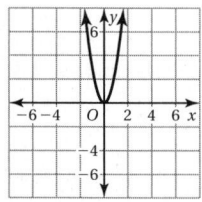

Both graphs open up, have the same vertex, $(0, 0)$, and the same axis of symmetry, $x = 0$. The graph of $y = 3x^2$ is narrower than the graph of $y = x^2$.

2.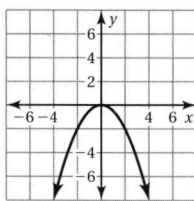

Both graphs have the same vertex, $(0, 0)$, and the same axis of symmetry, $x = 0$, but the graph of $y = -\dfrac{1}{2}x^2$ opens down. The graph of $y = -\dfrac{1}{2}x^2$ is a reflection in the x-axis and is wider than the graph of $y = x^2$.

3.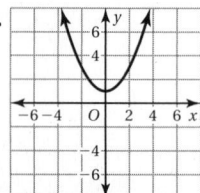

Both graphs open up and have the same axis of symmetry, $x = 0$. The graph of $y = \dfrac{1}{2}x^2 + 1$ is wider than and a translation 1 unit up of the graph of $y = x^2$.

4.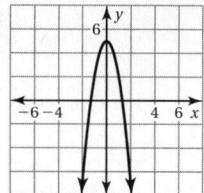

Both graphs have the axis of symmetry, $x = 0$, but the graph of $y = -3x^2 + 5$ opens down. The graph of $y = -3x^2 + 5$ is narrower than, a reflection in the x-axis, and a translation 5 units up of the graph of $y = x^2$.

5.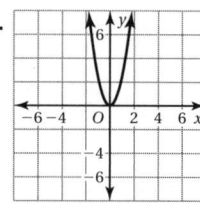

focus: $\left(0, \dfrac{1}{10}\right)$

6.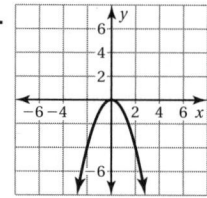

focus: $\left(0, -\dfrac{1}{4}\right)$

7.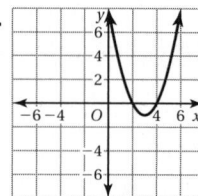

The domain is all real numbers. The range is all real numbers greater than or equal to -1.

8.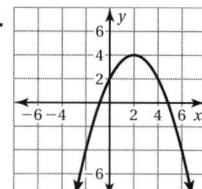

The domain is all real numbers. The range is all real numbers less than or equal to 4.

9.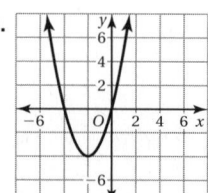

The graph of $y = (x + 2)^2 - 4$ is a translation 2 units to the left and 4 units down of the graph of $y = x^2$.

Answers

10.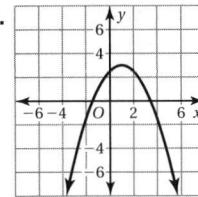

The graph of $y = -\frac{1}{2}(x-1)^2 + 3$ is a reflection in the x-axis, is wider than, and is a translation 1 unit to the right and 3 units up of the graph $y = x^2$.

11. $y = -\frac{1}{6}x^2$

12. a.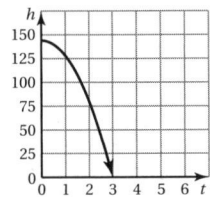

   b. 3 sec

13. quadratic; $y = -2x^2$

14. a. exponential b. $a = 2^m$ c. 64 in.

Alternative Assessment

1. a.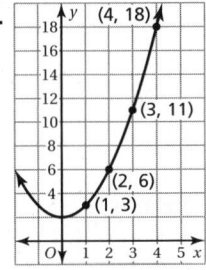

   b. quadratic function; The second differences are constant.

   c. $d = t^2 + 2$

   d. 38

   e. 9

   f. *Sample answer:* The time t represents the day and the distance d represents the number of miles you ride your bike each day.

2. a.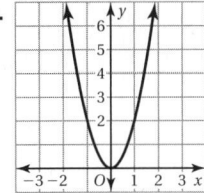

Both graphs have the same vertex, $(0, 0)$, and the same axis of symmetry, $x = 0$. The graph opens up and is narrower than the graph of $y = x^2$. The domain is all real numbers and the range is all real numbers greater than or equal to zero. The minimum value is 0.

b.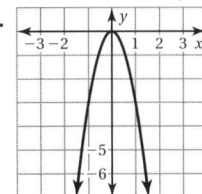

Both graphs have the same vertex, $(0, 0)$, and the same axis of symmetry, $x = 0$. The graph opens down and is narrower than the graph of $y = x^2$. The domain is all real numbers and the range is all real numbers less than or equal to zero. The maximum value is 0.

c.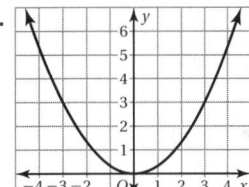

Both graphs have the same vertex, $(0, 0)$, and the same axis of symmetry, $x = 0$. The graph opens up and is wider than the graph of $y = x^2$. The domain is all real numbers and the range is all real numbers greater than or equal to zero. The minimum value is 0.

d.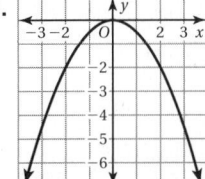

Both graphs have the same vertex, $(0, 0)$, and the same axis of symmetry, $x = 0$. The graph opens down and is wider than the graph of $y = x^2$. The domain is all real numbers and the range is all real numbers less than or equal to zero. The maximum value is 0.

Answers

e.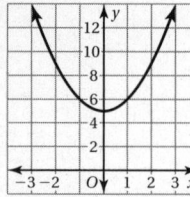

The vertex of the graph of $y = x^2 + 5$ is a vertical translation 5 units up of the vertex of the graph of $y = x^2$. Both graphs have the same axis of symmetry, $x = 0$. The domain is all real numbers and the range is all real numbers greater than or equal to five. The minimum value is 5.

f.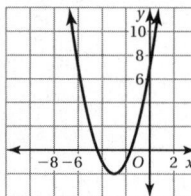

The vertex of the graph of $y = (x + 3)^2 - 2$ is a translation 3 units to the left and 2 units down of the vertex of the graph of $y = x^2$. The axis of symmetry is $x = -3$. The domain is all real numbers and the range is all real numbers greater than or equal to -2. The minimum value is -2.

g.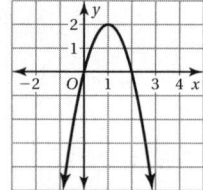

The vertex of the graph of $y = -2x^2 + 4x$ is a translation 1 unit to the right and 2 units up of the vertex of the graph of $y = x^2$. The axis of symmetry is $x = 1$. The domain is all real numbers and the range is all real numbers less than or equal to 2. The maximum value is 2.

h.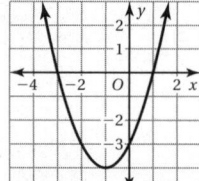

The vertex of the graph of $y = x^2 + 2x - 3$ is a translation 1 unit to the left and 4 units down of the vertex of the graph of $y = x^2$. The axis of symmetry is $x = -1$. The domain is all real numbers and the range is all real numbers greater than or equal to -4. The minimum value is -4.

Chapter 9

9.1–9.3 Quiz

1. $-4, 2$
2. -3

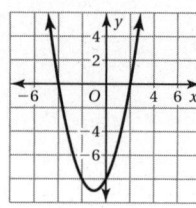 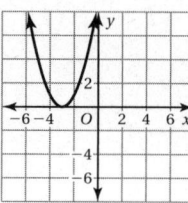

3. 0
4. $-2, 2$
5. no real solutions
6. $-7, 7$
7. $-1, 5$
8. -4
9. $x^2 + 2x + 1;\ (x + 1)^2$
10. $x^2 - 6x + 9;\ (x - 3)^2$
11. $2 \pm \sqrt{14}$
12. $-1, 11$
13. $-6, 2$
14. $-\dfrac{5}{2}, -\dfrac{1}{2}$
15. 1 sec
16. 4.07

9.4–9.5 Quiz

1. $-3, \dfrac{1}{2}$
2. $-\dfrac{5}{3}, 2$
3. no real solutions
4. $\dfrac{-2 \pm \sqrt{10}}{3}$
5. no real solutions
6. one real solution
7. $-9, 1$; *Sample answer:* Completing the square because the coefficient of the x^2-term is 1 and the coefficient on the *x*-term is even.
8. $-\dfrac{1}{2}, 3$; *Sample answer:* The quadratic formula because it can be used on any quadratic equation.
9. $-6, 3$; *Sample answer:* Factoring because the equation can be factored easily.
10. $-5, 0$; *Sample answer:* Factoring because the equation can be factored easily.
11. $(-3, -27), (12, 108)$
12. no real solutions
13. $(2, 1), (3, 2)$
14. $(0, 10), (2, 0)$
15. Hour 4
16. 2

Answers

Test A

1. $-5, 1$
2. 1

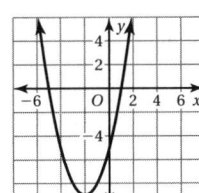
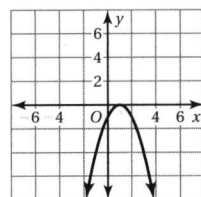

3. 0
4. no real solutions
5. $-2, 2$
6. $-8, 4$
7. 3 ft, 9 ft
8. $-4, 2$
9. $3 \pm \sqrt{2}$
10. $-6, 2$
11. $-\dfrac{9}{2}, -\dfrac{1}{2}$
12. $-1, 4$
13. $-1, -\dfrac{1}{2}$
14. $-\dfrac{7}{2}, \dfrac{7}{2}$
15. $1, \dfrac{1}{3}$
16. no real solutions
17. two real solutions
18. one real solution
19. no real solutions
20. $3 \pm \sqrt{5}$; Sample answer: Completing the square because the coefficient on the x^2-term is 1 and the coefficient on the x-term is even.
21. $-10, 3$; Sample answer: Factoring because the equation can be factored easily.
22. $-8, \dfrac{1}{2}$; Sample answer: The quadratic formula because it can be used on any quadratic equation.
23. $\pm 2\sqrt{2}$; Sample answer: Factoring because the equation can be factored easily.
24. $(9, 5), (-1, -5)$
25. $(-2, 4), (9, 81)$
26. no real solutions
27. $\left(1, \dfrac{1}{2}\right)$
28. $b = 12$
29. 4 sec
30. $6x^2 + 36x = 432$; 12 ft

Test B

1. $-3, 3$
2. $-1, 3$

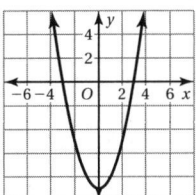
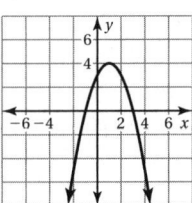

3. no real solutions
4. $\pm\sqrt{3}$
5. $-\dfrac{1}{3}, \dfrac{7}{3}$
6. $-5, 5$
7. 5 ft
8. $-3, 11$
9. $-6 \pm \sqrt{11}$
10. no real solutions
11. $-\dfrac{1}{2}, \dfrac{11}{2}$
12. $-4, 6$
13. $\dfrac{1}{3}$
14. $-3, \dfrac{1}{2}$
15. $\dfrac{4 \pm 2\sqrt{14}}{4}$
16. two real solutions
17. one real solution
18. two real solutions
19. no real solutions
20. $-2, 12$; Sample answer: Completing the square because the coefficient on the x^2-term is 1 and the coefficient on the x-term is even.
21. $\pm \dfrac{1}{4}$; Sample answer: Factoring because the equation can be factored easily.
22. $-1, \dfrac{5}{3}$; Sample answer: The quadratic formula because it can be used on any quadratic equation.
23. $-\dfrac{1}{2}, 0$; Sample answer: Factoring because the equation can be factored easily.
24. $(-4, -48), \left(-\dfrac{2}{3}, -\dfrac{4}{3}\right)$
25. no real solutions
26. $(-7, 7), (2, -2)$
27. no real solutions
28. $b = 3$
29. $\dfrac{1}{16}$ sec, 3 sec
30. $6x^2 + 48x = 768$; 16 ft

Answers

Alternative Assessment

1. **a.** 2 real solutions

   b. -4 and $-\dfrac{1}{2}$

   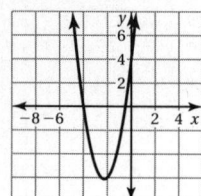

   c. *Sample answers:*

   Method 1: Solve by factoring
   $$2x^2 + 9x + 4 = 0$$
   $$(2x + 1)(x + 4) = 0$$
   $$2x + 1 = 0 \quad \text{or} \quad x + 4 = 0$$
   $$x = -\dfrac{1}{2} \quad \text{or} \quad x = -4$$

   Method 2: Solve by quadratic formula
   $$x = \dfrac{-(9) \pm \sqrt{(9)^2 - 4(2)(4)}}{2(2)}$$
   $$= \dfrac{-9 \pm \sqrt{49}}{4}$$
   $$= -\dfrac{9}{4} \pm \dfrac{7}{4}$$
   $$= -4 \text{ and } -\dfrac{1}{2}$$

   d. *Sample answer:* Solving by factoring because the equation can easily be factored.

2. **a.** 2 real solutions

   b. *Sample answers:*

   Method 1: Solve by completing the square
   $$x^2 - 4x - 14 = 0$$
   $$x^2 - 4x + 4 = 14 + 4$$
   $$(x - 2)^2 = 18$$
   $$x - 2 = \pm 3\sqrt{2}$$
   $$x = 2 \pm 3\sqrt{2}$$

   Method 2: Solve by quadratic formula
   $$x = \dfrac{-(-4) \pm \sqrt{(-14)^2 - 4(1)(-14)}}{2(1)}$$
   $$= \dfrac{4 \pm \sqrt{252}}{4}$$
   $$= 2 \pm 3\sqrt{2}$$

   c. *Sample answers:* Solving by completing the square because $a = 1$ and b is even.

3. **a.** 1 real solution

   b. 3

   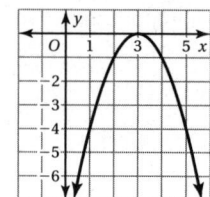

   c. *Sample answers:*

   Method 1: Solve by factoring
   $$-x^2 + 6x - 9 = 0$$
   $$-(x - 3)^2 = 0$$
   $$x - 3 = 0$$
   $$x = 3$$

   Method 2: Solve by quadratic formula
   $$x = \dfrac{-(6) \pm \sqrt{(6)^2 - 4(-1)(-9)}}{2(-1)}$$
   $$= \dfrac{-6 \pm 0}{-2}$$
   $$= 3$$

   d. *Sample answer:* Solving by factoring because the equation can easily be factored.

Answers

4. a. no real solutions

b. *Sample answers:*

Method 1: Solve by factoring
$$2x^2 + 18 = 0$$
$$2x^2 = -18$$
$$x^2 = -9$$
$$x = \pm\sqrt{-9}$$

no real solutions

Method 2: Solve by quadratic formula
$$x = \frac{-(0) \pm \sqrt{(0)^2 - 4(2)(18)}}{2(2)}$$
$$= \frac{0 \pm \sqrt{-144}}{4}$$

no real solutions

c. *Sample answer:* Solving by factoring because the equation can easily be factored.

5. The discriminant determines the number of real solutions of a quadratic equation. The x-intercepts are the real solutions to the equation. If the discriminant is positive, there are 2 real solutions and the graph intersects the x-axis twice. If the discriminant is zero, there is 1 real solution and the graph intersects the x-axis once. If the discriminant if negative, there are no real solutions and the graph does not intersect the x-axis.

Chapter 10

10.1–10.2 Quiz

1. $x \geq 0$ **2.** $x \geq -2$

3.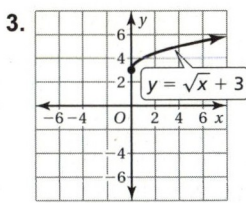

Domain: $x \geq 0$; Range: $y \geq 3$; The graph of $y = \sqrt{x} + 3$ is a vertical translation 3 units up of the graph of $y = \sqrt{x}$.

4.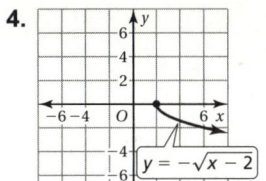

Domain: $x \geq 2$; Range: $y \leq 0$; The graph of $y = -\sqrt{x-2}$ is a reflection in the x-axis and a horizontal translation 2 units to the right of the graph of $y = \sqrt{x}$.

5. $x = 16$ **6.** $x = 144$ **7.** $x = 1$

8. $x = 37$ **9.** $x = \dfrac{1}{2}$ **10.** $x = 7$

11. $x = 4$ **12.** $x = 9$ **13.** 7.3 ft

10.3–10.4 Quiz

1. 26 ft   **2.** 8 in.   **3.** $4\sqrt{7}$ m

4. $2\sqrt{11}$ cm   **5.** yes   **6.** no

7. 5   **8.** 13   **9.** $\sqrt{34}$

10. $2\sqrt{10}$ **11.** yes   **12.** 67.1 ft

13. 181.1 ft   **14.** 130 ft   **15.** no

Test A

1.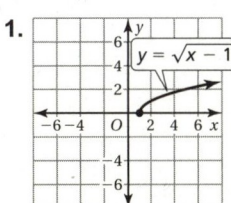

Domain: $x \geq 1$; Range: $y \geq 0$; The graph of $y = \sqrt{x-1}$ is a horizontal translation 1 unit to the right of the graph of $y = \sqrt{x}$.

2.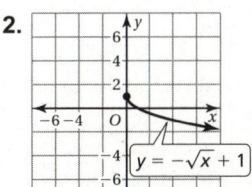

Domain: $x \geq 0$; Range: $y \leq 1$; The graph of $y = -\sqrt{x} + 1$ is a reflection in the x-axis and a vertical translation 1 unit up of the graph of $y = \sqrt{x}$.

Answers

3. $x = 81$ **4.** $x = 39$ **5.** $x = 18$

6. $x = 10$; $x = -2$ is extraneous.

7. $x = 4$ **8.** $x = 3$ **9.** $x = 1$

10. 7.5 yd   **11.** $2\sqrt{14}$ mm   **12.** yes

13. no   **14.** 17   **15.** 5

16. $\sqrt{65}$ **17.** 10   **18.** $3\sqrt{2}$

19. $\sqrt{82}$ **20.** $y \geq -2$

21. 144 ohms   **22.** 4.8 ft

23. a. 54.0 in.

  b. no; Because $a^2 + b^2 \neq c^2$, the triangle does not form a right triangle.

Test B

1.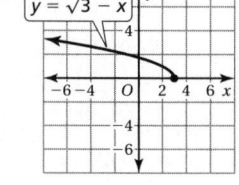

Domain: $x \leq 3$; Range: $y \geq 0$; The graph of $y = \sqrt{3 - x}$ is a reflection in the y-axis and a horizontal translation 3 units to the right of the graph of $y = \sqrt{x}$.

2.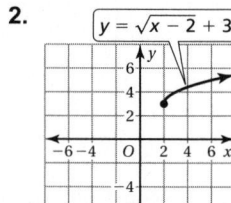

Domain: $x \geq 2$; Range: $y \geq 3$; The graph of $y = \sqrt{x - 2} + 3$ is a horizontal translation 2 units to the right and a vertical translation 3 units up of the graph of $y = \sqrt{x}$.

3. $x = 25$ **4.** $x = 64$ **5.** $x = 23$

6. $x = 2$; $x = -\dfrac{1}{2}$ is extraneous.

7. $x = 1$ **8.** $x = 3$ **9.** $x = 12$

10. 25 cm   **11.** $\sqrt{69}$ yd   **12.** no

13. yes   **14.** 13   **15.** $3\sqrt{10}$

16. $\sqrt{2}$ **17.** 10   **18.** $\sqrt{34}$

19. $2\sqrt{5}$ **20.** $y \leq 0$ **21.** 16 ohms

22. Kite A; Using the Pythagorean Theorem, the string length of Kite A is 37 feet and Kite B is 41 feet.

23. a. $10\sqrt{2}$ ft   **b.** $20\sqrt{2}$ ft

Alternative Assessment

1. a.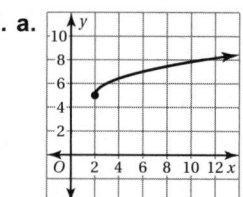

  b. The graph is a translation 2 units to the right and 5 units up of the graph of $y = \sqrt{x}$.

  c. domain: $x \geq 2$; range: $y \geq 5$

  d. $f(11) = 8$; $f(1)$ is undefined.

  e. $x = 6$; undefined

2. a. 300 yd

  b. 320 yd

  c. yes; The distance between the movie theater and the school is 500 yards. Because $300^2 + 400^2 = 500^2$, the triangle is a right triangle.

  d. no; Because $225^2 + 450^2 \neq 500^2$, the triangle is not a right triangle.

Chapter 11

11.1–11.3 Quiz

1. inverse   **2.** direct

3. $y = \dfrac{3}{4}x$

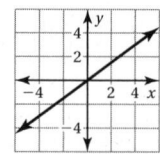

4. $x = 1$; $y = 2$; domain: all real numbers except 1; range: all real numbers except 2

Answers

5. $x = 0$; $y = -3$; domain: all real numbers except 0; range: all real numbers except -3

6. $f^{-1}(x) = \frac{1}{4}x + \frac{1}{4}$

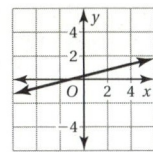

7. $f^{-1}(x) = \sqrt{x - 2}$

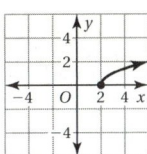

8. $\frac{t^2}{5}$; $t = 0$

9. $\frac{9}{w^5}$; $w = 0$

10. $\frac{a - 3}{a - 1}$; $a = -3$, $a = 1$

11. $\frac{b + 5}{b - 4}$; $b = -3$, $b = 4$

12. inverse variation; The product of the number of charities and the amount each charity receives is constant.

13. $\frac{4}{x + 1}$

11.4–11.7 Quiz

1. $\frac{2}{3z}$

2. $\frac{m + 3}{2m}$

3. $\frac{x - 1}{x - 6}$

4. $\frac{3}{2b(b - 5)}$

5. $p + 7$

6. $c - 7 + \frac{6}{c + 2}$

7. $y + 2$

8. $\frac{3x - 1}{(x + 1)^2}$

9. $-\frac{k + 2}{10k^2}$

10. $\frac{(m + 2)(m - 1)}{2(m + 3)(m - 3)}$

11. -1

12. $-4, -2$

13. 4

14. all real numbers except -3 and 3

15. $6 + \frac{36}{x - 2}$

16. 12 peanuts

Test A

1. $y = -6x$

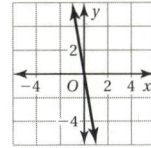

2. $y = \frac{2}{x}$

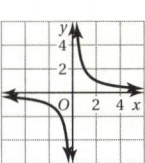

3. direct variation; The ratio of the total cost to the number of tickets is constant.

4.

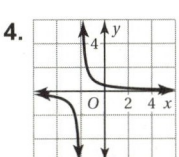

The graph is a translation 2 units to the left of $y = \frac{1}{x}$.

5.

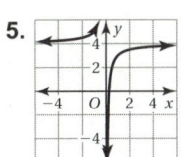

The graph is a reflection in the x-axis and a translation 4 units up of the graph of $y = \frac{1}{x}$.

6. $f^{-1}(x) = -\frac{1}{2}x + 2$

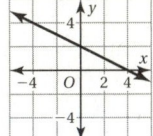

7. $f^{-1}(x) = \sqrt{x + 4}$

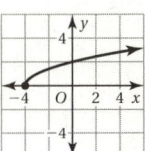

8. $\frac{3}{f}$

9. $\frac{2(a + 1)}{a - 2}$

10. $\frac{1}{2(u + 3)}$

11. $\frac{1}{g}$

12. $q + 6$

13. $e - 1 + \frac{4}{e - 3}$

14. $\frac{3x^2 - x + 4}{x + 2}$

15. $\frac{2c^2 + 3c + 4}{(c + 1)(c - 1)}$

16. -26

17. $2, 3$

18. 15

19. 3

Answers

20.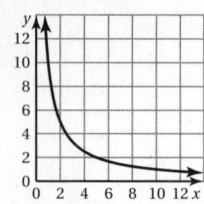

As the time increases, the number of songs per minute decreases.

21. 3.75 h

Test B

1. $y = -3x$ **2.** $y = \dfrac{12}{x}$

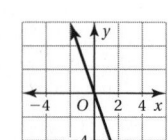 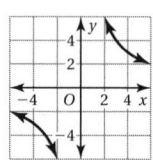

3. inverse variation; The product of the number of children and the cost per child is constant.

4.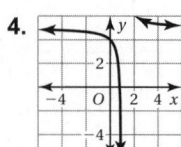

The graph is a translation 1 unit to the right and 5 units up of $y = \dfrac{1}{x}$.

5.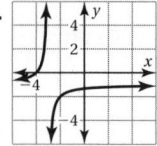

The graph is reflected about the x-axis and is a translation of 3 units left and 1 unit down of $y = \dfrac{1}{x}$.

6. $f^{-1}(x) = 2x - 4$ **7.** $f^{-1}(x) = \sqrt{-2x}$

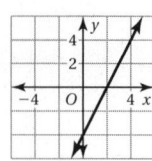 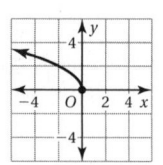

8. $\dfrac{2d}{5}$ **9.** $\dfrac{4x^2 - 4x + 1}{3(2x^2 + 5x - 2)}$

10. $\dfrac{4w}{3}$ **11.** $\dfrac{2n(n + 5)}{n - 5}$

12. $3k - 5$ **13.** $2y - 6 + \dfrac{12}{y + 3}$

14. $\dfrac{7x - 1}{12x}$ **15.** $\dfrac{(w + 3)(2w - 1)}{(2 - w)(w + 6)}$

16. 5 **17.** $-2, 5$ **18.** 6, 8 **19.** 12

20. a. $y = \dfrac{75}{t}$ **b.** $75 **c.** 7.5 h

21. $5\dfrac{1}{7}$ h

Alternative Assessment

1. a.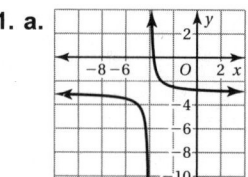

b. horizontal asymptote: $y = -3$; vertical asymptote: $x = -4$; domain: all real numbers except -4; range: all real numbers except -3.

c. $f^{-1}(x) = \dfrac{1}{x + 3} - 4$

d.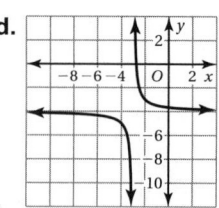

e. horizontal asymptote: $y = -4$; vertical asymptote: $x = -3$.

Answers

2. a. A; $\dfrac{x^6}{x-1}$; $x = 0$ and $x = 1$

b. *Sample answer:* $A \cdot B = x(x - 3)$;

$x = 0$ and $x = 1$; Common factors can easily divide out of both expressions.

c. *Sample answer:* $B \div C = \dfrac{(x-3)}{x(x+1)}$;

$x = 0$, $x = 1$, $x = -1$; After you multiply by the reciprocal, common factors can easily divide out of both expressions.

d. *Sample answer:*

$B + C = \dfrac{x^3 + x^2 - 5x + 3}{x^5}$;

$B - C = \dfrac{-x^3 + x^2 - 3x + 3}{x^5}$;

Each expression can easily be rewritten using the LCD.

e. $x = 1$

Chapter 12

12.1–12.4 Quiz

1. mean: 7; median: 7; mode: none

2. mean: 2; median: 2; modes: 0, 1, and 2

3. Joseph's times: mean: 37; range: 7; standard deviation: 2.38;

Daniel's times: mean: 38; range: 4; standard deviation: 1.53;

Joseph's times have a smaller mean, so his times are faster than Daniel's times. Daniel's times have a smaller standard deviation, so Joseph's times are more spread out.

4.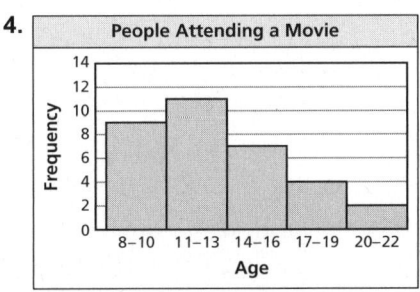

skewed right

5. a.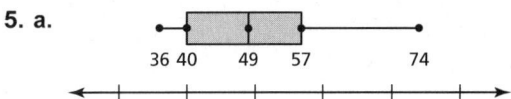

b. 17; The middle half of the points scored vary by no more than 17 points.

c. One-quarter of the games had a score of 40 points or less. One-half of the games had a score between 40 and 57 points. One-quarter of the games had a score of 57 points or higher.

12.5–12.8 Quiz

1. a. 2006   **b.** 47 male teachers   **c.** positive

2. a.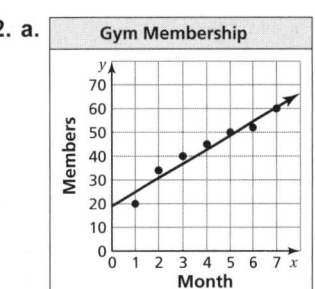

b. *Sample answer:* $y = 5.9x + 19$

c. Each month, the number of members increases by about 5.9 members.

d. *Sample answer:* about 90 members

3.

		Sports		
		Involved	Not involved	Total
Music Program	Involved	62	34	96
	Not involved	41	27	68
	Total	103	61	164

Sample answer: 164 students were surveyed. 103 students are involved in sports. 96 students are involved in the music program.

Test A

1. mean: 3; median: 3; mode: 3

2. mean: 41; median: 40; modes: 40 and 54

Answers

3. Sam's Sandwiches: mean: $6.50; range: $3; standard deviation: 0.96;

   Hugo's Hoagies: mean: $6; range: $4; standard deviation: 1.41;

   Hugo's Hoagies' prices have a smaller mean, so the prices are a little lower. Hugo's Hoagies' prices have a higher standard deviation so the prices are more spread out.

4. a.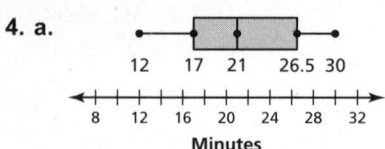

   b. range: 18; interquartile range: 9.5

   c. symmetric

5. *Sample answer:* line graph; shows changes over time

6. *Sample answer:* scatter plot; you want to compare two different data sets

7. *Sample answer:* bar graph: shows data in specific categories

8. a.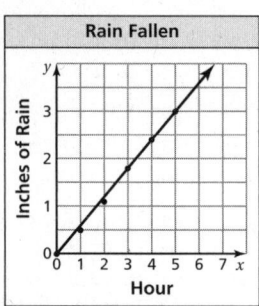

   b. *Sample answer:* $y = 0.6x$

   c. The amount of rainfall received each hour is about 0.6 inch.

   d. *Sample answer:* 4.8 in.

9. a. 12 students b. 52 students c. 60 students

   d.
		Breakfast		
		Ate	Skipped	Total
Lunch	Ate	40	12	52
	Skipped	8	0	8
	Total	48	12	60

   Sample answer: 48 students ate breakfast. 52 students ate lunch.

   e. 20%

Test B

1. mean: 16; median: 17.5; mode: 19

2. mean: −0.5; median: −0.5; mode: −1

3. Class A: mean: 88; range: 28; standard deviation: 7.96;

   Class B: mean: 85; range: 20; standard deviation: 6.68;

   The mean of Class A is higher, so the students scored better on the test. Class A has a higher standard deviation, so the scores are more spread out.

4. a.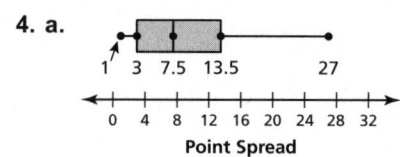

   b. range: 26; interquartile range: 10.5

   c. skewed right

   d. 27; The plot would be much smaller with a short right whisker.

5. *Sample answer:* circle graph; shows data as parts of a whole

6. *Sample answer:* bar graph; shows data in specific categories

7. a.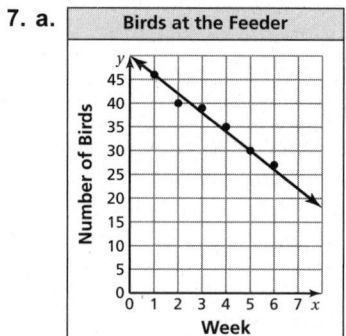

   b. *Sample answer:* $y = 50 - 4x$

   c. Each week the number of birds observed decreases by about 4.

   d. *Sample answer:* about 58 birds

Answers

8. a.

		Visited an Amusement Park		
		Yes	No	Total
Student	Boys	38	17	55
	Girls	29	25	54
	Total	67	42	109

b. *Sample answer:* 109 students were surveyed. 67 students visited an amusement park. 55 boys and 54 girls were surveyed.

c. boys: 69.1%; girls: 53.7%

Alternative Assessment

1. a. Data Set 1: bar graph; It shows the data in categories. A histogram cannot be used because the intervals are not the same size. To make the graph, put the number of boxes on the horizontal axis and the price on the vertical axis

Data Set 2: line graph; The data changes over time. To make the graph, put the months on the horizontal axis and the hours of daylight on the vertical axis. He can connect the points because the number of hours of daylight is continuously changing.

Data Set 3: circle graph; The type of car rented can be shown as parts of a whole. To make the graph, find the number of degrees for each car type by making a fraction from the number rented and the total of 560, and multiplying that fraction by 360°.

b. mean: 12.37; median: 12.25; mode: none; range: 9

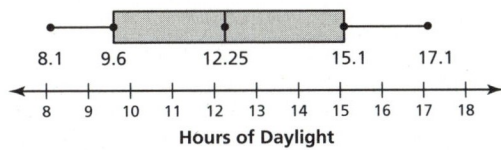

Sample answer: The hours of daylight seem to be evenly distributed.

2. a–b. *Sample answer:* When comparing length and weight, house cat is an outlier and the line of best fit of the other three cats is
$$w = 73\tfrac{1}{3}\ell - 403\tfrac{1}{3}.$$

When comparing length and speed, cheetah is an outlier and the line of best fit of the other three cats is $s = 2\ell + 27$.

When comparing weight and speed, cheetah is an outlier and the line of best fit of the other three cats is $s = 0.04w + 31.8$.

End-of-Course Test 1

1. $x = -6$ **2.** $c = 4$ **3.** 300 minutes

4. a. $A = \dfrac{1}{2}bh$ **b.** $h = \dfrac{2A}{b}$ **c.** 6 in.; 15.24 cm

5. slope: 3; y-intercept: -2

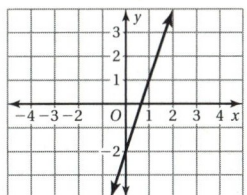

6. slope: $-\dfrac{1}{2}$; y-intercept: $\dfrac{3}{2}$

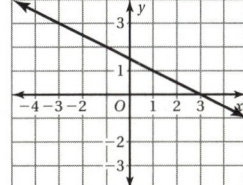

7. x-intercept: 4; y-intercept: 10; 4 adults can attend the play for $20 or 10 children can attend the play for $20.

8. $(-1, 1)$ **9.** no solution

10. $(-3, 0), (4, 7)$

11. 30 text messages and 25 pictures

12. $y = -\dfrac{2}{3}x + \dfrac{7}{3}$ **13.** $y = -2.5x + 6.5$

Answers

14. a. $y = \dfrac{9}{5}x + 32$

   b. $\dfrac{9}{5}$, or 1.8; 1°C corresponds to 1.8°F.

   c. 32; 0°C corresponds to 32°F.

15. domain: −2, −1, 0, 1, 2, 3; range: −3, −1, 1, 3, 5, 7

16. domain: −3, −2, −1, 0, 1, 2; range: 0, 2, 3, 4

17. a.

   b. discrete (You cannot buy part of a drink.)

   c. $y = \dfrac{3}{2}x$

   d. $4.50

18. $\dfrac{3}{2x^2 y}$ **19.** $\dfrac{4}{9a^4 b^2}$ **20.** 5 **21.** 16

22. $v^2 - 10v + 12$ **23.** $4z^2 + 12z + 9$

24. $\dfrac{x-3}{4x^2(x+2)}$ **25.** $\dfrac{g^2 + 3g - 12}{g(g-4)}$

26. $(s-8)(s-6)$ **27.** $b(b+7)(b-7)$

28. 13 ft

29. a. 42; 40; 40

   b.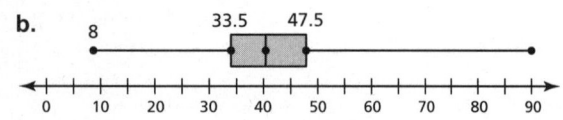

30. a–b.

   c. *Sample answer:* $y = 4x + 5$

   d. *Sample answer:* $21 per hour

31. Circle graph; A circle graph shows data as parts of a whole.

32. Line graph; A line graph shows how data changes over time.

33. $t - 3 \leq 7$ **34.** $4m > 12$

35. $x > -10$

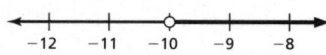

36. $x \leq 3$

37. $12x + 5 \geq 50$; $x \geq 3.75$; at least four CDs

38. $x = 6$ **39.** $a = -7, 2$

40. $r = -2 \pm 2\sqrt{5}$ **41.** $z = \dfrac{5}{4} \pm \dfrac{\sqrt{17}}{4}$

42. $x = 117$ **43.** $x = -3, 1$

44.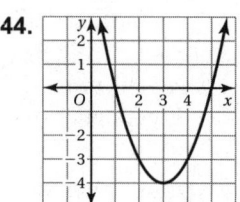

domain: all real numbers; range: $y \geq -4$

45.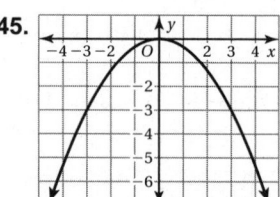

domain: all real numbers; range: $y \leq 0$

46.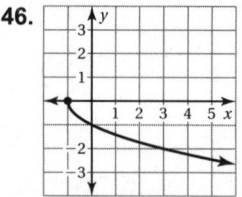

domain: $x \geq -1$; range: $y \leq 0$

Answers

47.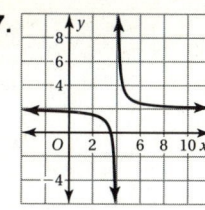

domain: $x \neq 4$; range: $y \neq 2$

End-of-Course Test 2

1. $r = -2.4$ **2.** $c = -2.5$ **3.** 145 minutes

4. a. $A = \frac{1}{2}bh$; $h = \frac{2A}{b}$

  b. 6.4 in.; 16.256 cm

5. slope: 1.5; y-intercept: 1

6. slope: $-\frac{3}{5}$; y-intercept: $\frac{1}{5}$

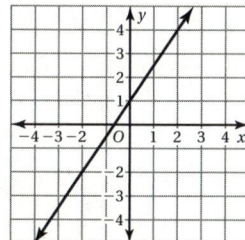 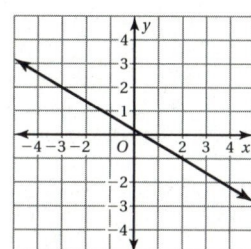

7. x-intercept: 6; y-intercept: 14; 6 adults can attend the play for $21 or 14 children can attend the play for $21.

8. $\left(-\frac{3}{2}, -\frac{1}{4}\right)$ **9.** no solution

10. $(-5, 0), (2, 14)$ **11.** 26 text messages

12. $y = -1.2x + 0.2$ **13.** $y = -2x + 7$

14. domain: $-2.5, -1, 0, 1.2, 2$;
range: $-3, -1, 1, 3, 5$

15. domain: $-1.5, -0.5, 0, 0.5, 1.5$;
range: $0, 2, 4$

16. volume in liters; it makes sense to have part of a liter, but it does not make sense to have a fraction of a car.

17. a.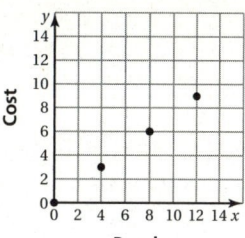

  b. discrete (You cannot buy part of a peach.)
  c. $y = 0.75x$ **d.** $4.50

18. $\dfrac{y^6}{x^8}$ **19.** $\dfrac{8}{3a^3b^4}$ **20.** 7   **21.** 10,000

22. $-6p^2 - 3p + 1$ **23.** $9n^2 - 6n + 1$

24. $\dfrac{3(c+3)}{c^2(c-3)}$ **25.** $\dfrac{u^2 + 6u - 3}{u(u+1)}$

26. $(a-9)(a+6)$ **27.** $f(f+7)^2$

28. 25 ft

29. a. 41.75; 40; 40; 83
  b.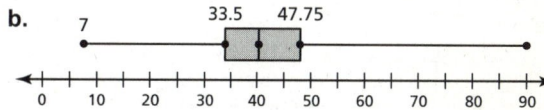

30. Circle graph; A circle graph shows data as parts of a whole.

31. Line graph; A line graph shows how data changes over time.

32. a.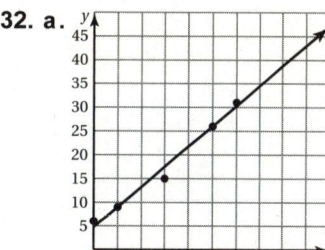

  b. Sample answer: $y = 4.25x + 4.75$
  c. Sample answer: about $22 per hour

33. $t - 3.2 \leq 7.5$ **34.** $\dfrac{4}{7}m > \dfrac{12}{5}$

35. $x > -6.8$

Answers

36. $x \leq \dfrac{2}{3}$

37. $12.50x + 5 \geq 50$; $x \geq 3.6$; at least four CDs

38. $x = -1$

39. $w = -5, 5$

40. $x = -9, 3$

41. no solution

42. $z = -1$

43. $y = -2, 10$

44.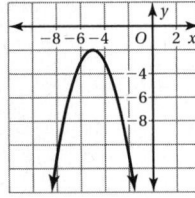

domain: all real numbers; range: $y \leq -2$

45.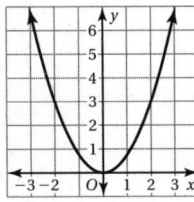

domain: all real numbers; range: $y \geq 0$

46.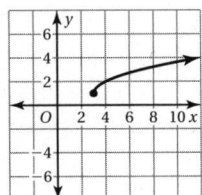

domain: $x \geq 3$; range: $y \geq 1$

47.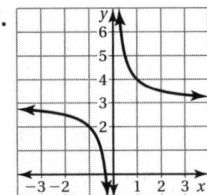

domain: $x \neq 0$; range: $y \neq 3$